Beyond the Law

Agribusiness and the Systemic Abuse of Animals Raised for Food or Food Production

by David J. Wolfson*

---

*David J. Wolfson is an attorney at Milbank, Tweed, Hadley & McCloy in New York City. The first edition of this article was published by Archimedian Press/Coalition for Non-Violent Food and reprinted in *Animal Law*, Volume 2.

BEYOND THE LAW
Agribusiness and the Systemic Abuse of
Animals Raised for Food or Food Production
Written by David J. Wolfson

All rights reserved. No part of this book may be reproduced in any form or by any means, electronic or mechanical, including photocopying, recording, or by any information storage and retrieval system without written permission from the publisher, except for the inclusion of brief quotations in a review.

Cover Design: John B. Moretti / Design Production: Laura A. Moretti

Copyright © 1999 by Farm Sanctuary, Inc.
Printed in the United States of America
ISBN: 0-9656377-1-9 $2.50   Softcover

All photos, unless otherwise copyrighted, are the property of Farm Sanctuary, Inc.

Beyond the Law was first published in 1996 by Archimedian Press, a project of the late Henry Spira and Animal Rights International. In his Publisher's Note, Henry wrote, "*Beyond the Law* is designed to bring to the public's attention the world of misery in which more than eight billion farm animals live, suffer and die in violence each year in the USA." Farm Sanctuary thanks Henry Spira for his work on behalf of farm animals and Animal Rights International for helping to fund this updated version of *Beyond the Law*.

"...if one person is unkind to an animal, it is considered to be cruelty, but where a lot of people are unkind to animals, especially in the name of commerce, the cruelty is condoned and, once sums of money are at stake, will be defended to the last by otherwise intelligent people."

Ruth Harrison, *Animal Machines*

PREFACE

In recent decades, farm animals in America have been subjected to increasingly inhumane conditions. Meanwhile, a growing number of consumers are voicing ethical concerns about methods used to produce meat, milk, and eggs.

Many practices common in animal agriculture—such as intensive confinement housing—face widespread public opposition. A 1995 poll conducted by Opinion Research Corporation for Animal Rights International found that at least 90% of Americans disapprove of confinement methods used in the production of eggs, veal, and pork. Confronted with growing ethical resistance to inhumane farming practices, animal agriculture has campaigned to strip legal protection from farm animals.

In *Beyond the Law*, David J. Wolfson, Esq. explains how cruel farming practices are being excluded from legal censure throughout the United States. He discusses how agribusiness has enacted legislation, giving itself the authority to regulate how farm animals can be treated. Wolfson writes,

> ...thirty states have enacted laws that create a legal realm whereby certain acts, no matter how cruel, are outside the reach of anticruelty statutes as long as the acts are deemed "accepted," "common," "customary," or "normal" farming practices. These statutes have given the farming community the power to define cruelty to animals in their care.

He contrasts this trend with other industrialized countries who are passing stronger laws to protect farm animals. *Beyond the Law* cogently illuminates how an influential special interest has manipulated the legislative process to advance legal policies which offend public sentiments. It is a wakeup call for compassionate citizens throughout the U.S.

Beyond the Law

> [The argument was presented that] any practice which accorded with the norm in modern farming or slaughter practices was thereby acceptable and not to be criticized as cruel. I cannot accept this approach.... To do so would be to hand the decision as to what is cruel to the food industry completely, moved as it must be by economic as well as animal welfare considerations.*

---

*Mr. Justice Bell, McLibel Verdict, Section 8, June 19, 1997 [hereinafter McLIBEL VERDICT].

Above: A calf is burned with a hot-iron brand.

I. INTRODUCTION[1]

Since the early nineteenth century, Western society has enacted laws to protect animals from cruelty. While such laws were originally intended to protect animals such as cows, sheep, and horses, they have generally evolved to cover all domestic animals, including dogs and cats. In recent years, however, a large number of U.S. states have amended their anticruelty laws. Today, the majority of U.S. states prohibit, at least in part, the application of their anticruelty statutes to farm animals.

Specifically, 30 states have enacted laws that create a legal realm whereby certain acts, no matter how cruel, are outside the reach of anticruelty statutes as long as the acts are deemed "accepted," "common," "customary," or "normal" farming practices. These statutes have given the farming community the power to define cruelty to animals in their care. Similarly, certain states' anticruelty statutes also exclude poultry, which represent an estimated 95% of the approximately eight billion farm animals slaughtered annually.

Eighteen states in the last ten years have amended their statutes to exempt "accepted," "common," "customary," or "normal" farming practices; since 1994, Connecticut, Idaho, Iowa, Michigan, New Jersey North Carolina, and Wyoming joined this trend, moving the states that exempt such farming practices from the minority to the majority. Such amendments indicate that current methods of farming encompass practices that might have been considered illegal prior to the amendments. Consequently, it was necessary to amend the anticruelty statutes to allow such practices to continue. Given that there is also no federal law applicable to the treatment of animals raised for food or food production while on the farm, such animals within these states now have no legal protection from institutionalized cruelty.

This booklet will examine both federal and state protection of animals raised for food or food production. Part II demonstrates how these animals receive absolutely no federal protection while on the farm and extremely limited federal protection during transport and slaughter. It will also discuss the sole source of legal protection for such animals on the farm: state anticruelty statutes, which themselves offer questionable

---

[1] I would like to thank the following individuals for their invaluable input and criticism: Elinor Molbegott, Kathrin Wanner, Gene Bauston, Dr. Andrew Rowan, Dr. Louise A. S. Murray, Lisa Weisberg, Steven Wise, Martha Fineman, Merritt Clifton, Peter Singer, Peter Stevenson, and Laura A. Moretti. In particular, I would like to recognize the late Henry Spira, who encouraged me to write this booklet and who provided essential insight and input. As with many achievements in the animal protection movement, this booklet would simply have been impossible without his existence.

protection. Part III briefly discusses, "accepted," "common," "customary," or "normal" farming practices.[2] Part IV documents how the amendments to anticruelty laws mentioned above place animals raised for food or food production beyond the law's reach in the majority of states.

Part V offers evidence that the law in the United States fails to protect animals raised for food or food production from cruel treatment through a comparison with legal developments in Western Europe. Not only are many customary farming practices in this country cruel, but many Western European countries have recognized the cruelty in such practices and are attempting to remedy it. Furthermore, even if present customary farming practices are not perceived as cruel, statutory exemptions creates a realm whereby cruel farming practices can be developed without fear of sanction. The contrast is stark: the United States alters the law to allow cruel farming practices while Western European countries are banning cruel farming practices. Part VI concludes with an outline for reform.

The main purpose of this booklet is not remedial, but rather to present the realities of the current system. Although many people may have the impression that laws prevent domestic animals—the vast majority of which are animals raised for food or food production—from being treated in a cruel manner, the reality is that more such animals are now being abused than ever before in the history of the United States.

II. LEGAL PROTECTION FOR ANIMALS RAISED FOR FOOD OR FOOD PRODUCTION

The worst sin toward our fellow creatures is not to hate them, but to be indifferent to them. That's the essence of inhumanity.[3]

Although animals lack legal standing, they are protected in small

---

[2]For the purpose of brevity, the term "customary farming practices" will be used to refer to "accepted," "common," "customary," and "normal" farming practices.
[3]George Bernard Shaw, *The Devil's Disciple*, in THE EXTENDED CIRCLE: A COMMONPLACE BOOK OF ANIMAL RIGHTS 325 (J. Wynne-Tyson ed.,1985) [hereinafter THE EXTENDED CIRCLE].

part by both state and federal statutes.[4] In terms of animals raised for food or food production, three separate areas of activity are in need of protection: the treatment of the animal while being reared, the transport of the animal to the slaughterhouse or stockyard, and the slaughter itself.

[4]For an argument toward extending such standing to animals, see Laurence Tribe, Ways Not To Think About Plastic Trees, 83 YALE L.J. 1315, 1341 (1974):

> At a minimum we must begin to extricate our nature-regarding impulses from the conceptually oppressive sphere of human want satisfaction, by encouraging the elaboration of perceived obligations to plant and animal life...in terms that do not falsify such perceptions from the very beginning by insistent reference to human interests.... And legislation might be enacted to permit the bringing of claims directly on behalf of natural objects without imposing the requirement that such claims be couched in terms of interference with human use.

For a related proposal suggesting the appointment of guardians or trustees for animals, see, Christopher Stone, Should Trees Have Standing?—Toward Legal Rights for Natural Objects, 45 S. CAL. L. REV. 450, 458-59 (1972). Our legal system has recognized many entities other than individual human beings: for example, churches, partnerships, corporations, and unions.

Above: Chickens raised for meat are packed by the thousands in warehouses.

A. Federal Protection

One of the most important pieces of federal legislation in recent years relating to animals is the Animal Welfare Act, which generally applies to animals used in research and exhibitions, and commercial breeders of dogs and cats sold for research and the pet trade.[5] In 1985, the Improved Standards for Laboratory Animals Act amended the Animal Welfare Act.[6] Among other things, the amendment required the establishment of institutional animal care and use committees, and the provision by the United States Department of Agriculture (USDA) of additional standards for the care of animals used in research.[7] However, the Animal Welfare Act does not apply to animals raised for food or food production, and consequently, is irrelevant to the issue at hand.[8]

By contrast, the Twenty-Eight Hour Law of 1877 does apply to animals raised for food or food production.[9] This law, which was repealed and reenacted in an amended form in 1994, provides that animals cannot be transported across state lines for more than 28 hours by a "rail carrier, express carrier, or common carrier (except by air or water)" without being unloaded for at least five hours of rest, watering, and feeding.[10] The statute does not apply to animals transported in a vehicle or vessel in which the animals have food, water, space and an opportunity to rest. Furthermore, sheep may be confined for an additional eight consecutive hours when the 28 hour period of confinement ends at night, and animals may be confined for 36 consecutive hours upon the request of the owner or the person having custody of the animals.[11]

While it is arguable whether 28 hours is a humane time limit when compared with the 15 hour time limit in Britain and the eight hour time limit for standard vehicles in the European Community, the law does not cover transport by air or water and transport within a state. Most notably, the law is rarely enforced by the Attorney General, and even if a

---

[5] 7 U.S.C. §§ 2131-2159 (1994).
[6] Id. 2143.
[7] Id.
[8] Section 2132(g) of the Animal Welfare Act reads: "[T]he term "animal"... excludes...farm animals, such as, but not limited to livestock or poultry used or intended for use as food or fiber, or livestock or poultry used or intended for improving animal nutrition, breeding, management, production, efficiency, or for improving the quality of food or fiber." Animal in the 1985 Amendment is similarly defined. 7 U.S.C. § 2132(g) (1988).
[9] 49 U.S.C. § 80502 (1995).
[10] Id.
[11] Id.

conviction occurs, the maximum penalty is only $500.[12] It is also questionable whether the law applies to trucks.[13]

Finally, the Humane Slaughter Act[14] requires that livestock slaughter "be carried out only by humane methods" to prevent "needless suffering".[15] Additionally, regulations enacted pursuant to the Humane Slaughter Act of 1978 forbid the dragging of conscious non-ambulatory animals.[16] However, these statutes apply only to slaughterhouses under Federal meat inspection, exclude poultry, and possess an important exemption for ritual slaughter.[17] The statutes do not cover state inspected and small custom exempt slaughterhouses. Ultimately, it is difficult to ascertain the effectiveness of the statute; there is insufficient enforcement and the slaughter-houses are off-limits to the general public.

In sum, no federal law regulates the first area of concern-how animals raised for food or food production are treated on the farm while being reared. Similarly, a limited, arguably inhumane, and largely unenforced federal law applies to the interstate transport of animals to sale or slaughter. The law applicable to actual slaughter is also problematic. It is therefore necessary to turn to the states in hope of further legal protection.

B. State Protection

All states in the United States have anticruelty laws, and certain states have laws pertaining to the transport and slaughter of animals. Indeed, the first statutory protection for animals was created in the United States. In 1641, the Puritans of Massachusetts Bay Colony voted to print their first legal code, "The Body of Liberties", which included Liberty 92 forbidding cruelty to animals: "No man shall exercise any Tiranny or Cruelty towards

---

[12]Id.
[13]As this paper will discuss, most states have transportation laws of their own, but many have time limits which are the same as, or exceed, those provided for under this law. ANIMAL WELFARE INSTITUTE, *ANIMALS AND THEIR LEGAL RIGHTS: A SUMMARY OF AMERICAN LAWS FROM 1641-1990* 50 (1990) [hereinafter ANIMAL WELFARE INSTITUTE].
[14]7 U.S.C. §§ 1901-1906 (1994); 9 C.F.R. § 301.2(qq) (1999). For an excellent discussion of the problems related to this statute, see, GAIL A. EISNITZ, *SLAUGHTERHOUSE* (1997).
[15]Id. § 1901
[16]9 C.F.R. § 313.2 (1993).
[17]Recent developments have significantly reduced the problems related to the "shackle and hoist" method of slaughter previously associated with ritual slaughter. Such developments are in large part due to the initiative of Henry Spira and Temple Grandin and the application of her upright restraining system.

any brute Creature which are usually kept for man's use."[18]

Other than this early law, there was no specific legislation for approximately two hundred years, although it was possible to prosecute cruelties under the common law as "nuisances." As far as can be determined, the first anticruelty law in the United States was enacted in 1821 by the Maine Legislature.[19] The law was limited to horses and cattle, and protected these species from being "cruelly beat."[20] Maine was followed by New York in 1829 with a law applied to horses, ox, cattle and sheep belonging to another person; Massachusetts in 1835; and Connecticut and Wisconsin in 1838.[21]

Historically, since the first agrarian societies domesticated animals for human use, farm animals have been viewed as personal property to be disposed of as the owner wished. Only when animals gained economic value did the law prohibit the interference with such animals by someone other than the owner.[22] For example, destruction of livestock became a crime. Dogs, however, were so "undeserving of societal concern that not only did the criminal system not protect dogs, but special statutes were passed to preclude a dog owner from seeking recourse for harm to the animal."[23] Ironically, the situation today is somewhat reversed; dogs and cats are granted some legal protection, but many state anticruelty statutes have been amended to prevent cruelty charges being brought for customary methods of food production.

The laws passed in the late 1800s are the core of today's legislation. It has been argued that many courts and authors were uncomfortable with criminal laws being based solely on the welfare of animals.[24] Consequently, anticruelty laws were justified on the ground that acts of cruelty dulled humanitarian feelings.[25] Such laws were intended primarily to protect humanity and society, rather than the animals themselves. Thus, the legal duty to animals has been perceived as somewhat indirect, based on "the proposition that we have no duty directly to animals; rather, animals are a sort of medium through which we may either succeed or fail to discharge those direct duties we owe to non-animals, either ourselves, other human

---

[18] ANIMAL WELFARE INSTITUTE, supra note 13, at 1.
[19] David Favre and Viven Tsang, The Development of Anticruelty Laws During the 1800's, 1 DET.C.L. REV. 1, 8 (1993).
[20] Id.
[21] Id. at 9; ANIMAL WELFARE INSTITUTE, supra note 13, at 2-3
[22] DAVID FAVRE & M. LORING, ANIMAL LAW 122 (1983).
[23] Id. See also, Charles Friend, Animal Cruelty Laws: The Case for Reform, 8 U. RICH. L. REV. 201 (1974).
[24] Favre & Loring, supra note 22, at 122.
[25] Id. See, Richard F. McCarthy & Richard E. Bennett, Statutory Protection for Farm Animals, 3 PACE ENVTL. L. REV. 229, 235 (1986).

beings, or, as on some views, God."[26] Interestingly, French law viewed this philosophy so literally that animal cruelty was only a crime when the cruel act occurred in public so as to affect human observers.[27]

Such views are still prevalent in the interpretation of today's anticruelty laws. As a court noted in 1981: "These [anticruelty] statutes are 'directed against acts which may be thought to dull humanitarian feelings and to corrupt morals of those who observe or have knowledge of those acts.'"[28] Indeed, a hundred years earlier, the Massachusetts Supreme Judicial Court stated that such statutes defined an offense not against the "rights of animals that are in a sense protected by it. The offense is against public morals, which the commission of cruel and barbarous acts tend to corrupt."[29]

With the exception of the limited federal laws that apply to transport and slaughter previously discussed, the sole protection from unnecessary suffering and cruel treatment for animals raised for food or food production falls within state criminal statutes.[30] While every state has an anticruelty law that forbids cruelty to animals in general, they vary significantly in degree and coverage. A brief survey of such laws leads to the following generalities.

---

[26]Tom Regan, *THE CASE FOR ANIMAL RIGHTS* 150 (1983). See also Steven Wise, Of Farm Animals and Justice, 3 PACE ENVTL. L. REV. 191, 205 (1986).
[27]Wise, supra note 26, at 205.
[28]Knox v. Massachusetts SPCA, 425 N.E.2d 393, 396 (Mass. 1981) (citing Commonwealth v. Higgins. 277 Mass. 191, 194 (1931)).
[29]Commonwealth v. Turner, 14 N.E. 130, 131-132 (1877).
[30]Attempts to invoke the civil process to protect animals have been largely unsuccessful. Standing problems exist when individuals assert the rights of third parties, and those who have standing—the owners of the "property"—nearly always have no desire to participate and are usually inflicting the "injury." Additionally, even if a private citizen attempts to halt a practice and can avoid this problem, the matter may be dismissed upon the traditional equitable ground that the court will not enjoin a criminal act. See, Animal Legal Defense Fund v. Provimi Veal Corp., 626 F. Supp. 278 (D. Mass. 1986): "An ALDF victory in this action would have an unmistakable effect: to enforce by means of an injunction obtained in a private lawsuit, a criminal statute enforceable only by public prosecutors . . .". Wise, supra note 26, at 217-218. An exception to this statement can be found in the recent case *Farm Sanctuary v. Department of Food and Agriculture*, Second Appellate District, Court of Appeal California (April 22, 1998, at 7), where it was held that Farm Sanctuary, a nonprofit organization that promotes the humane treatment of farm animals, could challenge a ritualistic slaughter regulation because, "unless an organization like Farm Sanctuary is permitted to challenge the [Department of Food and Agriculture's] rulemaking authority, the ritualistic slaughter regulation will be immune from review ... As one court has observed: 'Where [a statute] is expressly motivated by considerations of humaneness toward animals, who are uniquely incapable of defending their own interests in court, it strikes us as eminently logical to allow groups specifically concerned with animal welfare to invoke the aid of the courts in enforcing the statute.'"

On the Farm

Nineteen states and the District of Columbia prohibit both depriving an animal of "necessary sustenance" and failing to provide "food, water and shelter."[31] Several states require the provision of "necessary sustenance" without further reference to food, water, shelter,

and the application of this phrase varies from state to state.[32] "Approximately half the state statutes require shelter without qualifying phrases, but most states require the failure to provide shelter to be proven to be intentional or cruel."[33] Finally, nearly half the states have laws which stipulate that cruelty to animals is an offense only if committed "willfully", "maliciously" or "cruelly."[34] As will be discussed below, many

---

[31] ANIMAL WELFARE INSTITUTE, supra note 13, at 7-10.
[32] Id.
[33] Id. at 9.
[34] Id. at 7.

Above: Male chicks, discards of the egg industry, suffocate in dumpsters.

states mandate such requirements and then prohibit the application of the anticruelty statute to customary farming practices.

Transport

Federal law only applies to the interstate transport of animals, not the transport of animals within a state. The anticruelty laws or transportation laws of most states and the District of Columbia also require that transport of animals be conducted in a humane manner.[35] Nebraska and Nevada, however, specifically exempt animals raised for food or food production from their transportation laws, and some other states exempt them generally.[36] The laws that do exist are very brief. The following is a typical example of a transportation statute: "[I]f any person shall carry, or cause to be carried by hand or in or upon any vehicle or other conveyance, any creature in a cruel or inhumane manner, he shall be guilty of a misdemeanor."[37] Oregon has a specific provision whereby its anticruelty law does not apply to the "transport of livestock...or commercially grown poultry" unless there has been gross negligence.[38]

Furthermore, the fines for a breach of such transport laws are small: $1000 in Connecticut (or $100 for cruelty to poultry),[39] $400 in South Carolina,[40] $1000 in New Jersey,[41] and $150 in Washington.[42]

Similarly, the time limits for transporting animals without food, water, and rest are problematic. The shortest maximum time period an animal can be transported without food, water and rest is 18 hours for trucks in Vermont,[43] with many states allowing 28 hours for railroad and trucks.[44] States often allow 36 hours, if requested, for both railroad and

---

[35]Id. at 11.
[36]NEB. REV. STAT. § 28-1013(6) (Supp. 1994); NEV. REV. STAT. ANN. § 574.200(6) (Michie 1994).
[37]MISS. CODE ANN. § 97-41-5 (1972). ANIMAL WELFARE INSTITUTE, supra note 13, at 11.
[38]OR. REV. STAT. § 167.335 (1995).
[39]CONN. GEN. STAT. ANN. § 53-249 (1998).
[40]S. C. CODE ANN. §§ 47-1-40, 47-1-50 (Law. Co-op. Supp 1993).
[41]N.J. STAT. ANN. § 4:22-17 (1998).
[42]WASH. REV. CODE §§§ 16.52.070, 16.52.080, 16.52.165 (1994).
[43]VT. STAT. ANN. tit. 13, § 382 (Supp 1994).
[44]CONN. GEN. STAT. ANN. § 53-252 (West 1994), FLA. STAT. ANN. § 828.14 (West 1994), ILL. ANN. STAT. 510 ILCS 70/7 (Smith-Hurd 1993), MASS. ANN. LAWS. ch. 272 § 81 (Law Co-op Supp 1994), R.I. GEN. LAWS § 4-1-17 (1987), VA. CODE ANN. § 3.1-796.69 (1994), VT. STAT. ANN. tit. 13, § 382 (Supp 1994).

trucks,[45] and Washington allows a total of two days for the transport of animals without food, water, or rest on the railroad. A breach of this law can result in a fine of up to $1000.[46] The average fine for breach of state transportation laws is approximately $500.

A limited number of states also have a separate anticruelty law which pertains to the transportation of poultry.[47] In Pennsylvania, "it shall not be deemed cruel or inhumane to transport live poultry in crates so long as not more than fifteen pounds of live poultry are allocated to each cubic foot of space in the crate."[48] This is approximately four birds per cubic foot.

Slaughter

As described above, the limited protection of federal law in relation to slaughter necessitates that each state must pass its own humane slaughter law if all animals raised for food are to be protected from inhumane treatment.[49] Presently, 27 states have enacted humane slaughter laws.[50] Nine of these do not prohibit what is generally

---

[45]CAL. FOOD AND AGRIC. CODE § 16908 (West 1986), ME. REV. STAT. ANN. tit. 7, § 3981 (1989), MICH. COMP. LAWS ANN. § 28.246 (West 1991), Minn. Stat. Ann. § 343.24 (West 1991), NEV. REV. STAT. ANN. § 705.090 (Michie 1993), N.Y. AGRIC. AND MKTS. LAW § 359 (McKinney 1991), OHIO REV. CODE ANN. § 959.13 (Baldwin 1988), S. C. CODE ANN. § 47-1-90 (Law. Co-op. Supp. 1992).
[46]WASH. REV. CODE § 81.56.120 (1992).
[47]CONN. GEN. STAT. ANN. § 53-249 (West 1985), PA. STAT. ANN. tit. 18, § 5511(e) (Supp 1994), R.I. GEN. LAWS § 4-1-7 (1987), VT. STAT. ANN. tit. 13 § 352(a)(10) (Supp 1994), WIS. STAT. ANN. § 134.52 (West 1989).
[48]PA. STAT. ANN. tit. 18, § 5511(e) (Supp 1995).
[49]ANIMAL WELFARE INSTITUTE, supra note 13, at 57.
[50]ARIZ. REV. STAT. ANN. §§ 3-2016, 3-2017 (1995), CAL. FOOD & AGRIC. CODE § 19501 (West Supp 1993), COLO. REV. STAT. § 35-33-203 (Supp 92), CONN. GEN. STAT. ANN. § 22-272a (West 1985), FLA. STAT. ANN. § 828.22 (West 1994), GA. CODE ANN. § 26-2-110.1 (1982), HAW. REV. STAT. § 159-21 (1994), ILL. ANN. STAT. § 510 ILCS 75/1 (Smith-Hurd 1993), IND. CODE ANN. § 42-16-6 (Burns 1994), IOWA CODE ANN. § 189A.18 (West 1985), KAN. STAT. ANN. § 47-1401 (1993), MD. ANN. CODE art. 27 § 333B (1992), MASS. ANN. LAWS ch. 94, § 139D (Law. Co-op. 1985), MICH. COMP. LAWS ANN. § 12.484 (West 1989), MINN. STAT. ANN. § 31.591 (West 1980), N. H. REV. STAT. ANN. § 427:33 (1991), OHIO REV. CODE ANN. § 945.01 (Baldwin 1988), OKLA. STAT. ANN. tit. 2, § 6-183 (West 1993), OR. REV. STAT. ANN. § 603.065 (1988), PA. STAT. ANN. tit. 3, § 451.52 (Purdon Supp 1994), R.I. GEN. LAWS §§ 4-17-3, 4-17-4 (1987), S.D. CODIFIED LAWS ANN. § 39-5-23.2 (1985), UTAH CODE ANN. § 4-32-7(8)(b) (Supp 1994), VT. STAT. ANN. tit. 6, § 3131 (1988), WASH. REV. CODE § 16.50.120 (1992), W. VA. CODE § 19-2E-1 (1993), and WIS. STAT. ANN. § 95.80 (West 1990).

recognized as an inhumane method of stunning before slaughter (the manually operated sledgehammer),[51] and four have not even charged an official or department with the enforcement of the law.[52]

Moreover, 15 states have designated the State Department of Agriculture or the Board of Agriculture to be in charge of enforcement.[53] It must be recognized that the primary purpose of such agencies is not animal well-being. Finally, the fines for breach of a law average at about $500 with certain states limiting the fine to $100[54] or not specifying a penalty.[55]

[51]States that allow the manually-operated sledgehammer method of stunning are: Arizona, California, New Hampshire, Ohio, Oklahoma, South Dakota, Utah, West Virginia, and Wisconsin. ANIMAL WELFARE INSTITUTE. supra note 13, at 61.
[52]Georgia, Kansas, Maryland and Ohio. Id.
[53]California, Colorado, Illinois, Iowa, Michigan, Minnesota, New Hampshire, Oklahoma, Oregon, Pennsylvania, Utah, Vermont, Washington, West Virginia and Wisconsin. Id.
[54]Iowa, Michigan, Pennsylvania, Vermont, West Virginia, Wisconsin all limit the fine for animal cruelty. Id.
[55]Georgia. Id.

Above: For birthing, sows are confined in "farrowing" stalls.

Problems Related to the Enforcement of State Anticruelty Laws

While the above may seem to provide a limited degree of legal protection to animals raised for food or food production from unnecessary suffering and cruel treatment, any legal protection is curtailed by the following factors: First, it is important to note what is not protected as well as what is. "Provisions for adequate exercise, space, light, ventilation, and clean living conditions for confined animals are important but infrequent requirements of state anticruelty laws."[56] For example, light is a requirement in only the Washington[57] and Puerto Rico statutes.[58] Only Maine and Wisconsin refer to clean living conditions. Furthermore, the statutes that do mention such rights are vaguely worded.

Second, the statutes are frequently drafted in exceedingly general terms with discretion left to the court to exclude certain animals, or they specifically exclude certain animals, such as fowl;[59] in 1996, 7.7 billion chickens, turkeys and ducks were killed in the United States. Third, many state statutes require that the prosecution demonstrate a mental state of the defendant that may be hard to prove.[60] Finally, most laws are not effectively enforced, and enforcement is largely directed at dogs, cats, and horses rather than animals raised for food or food production.[61]

With regard to this final point, "[t]he enforcement of these criminal statutes is typically left to a public prosecutorial agency, itself overwhelmed by human problems, or to an overburdened private Society for the Prevention of Cruelty to Animals (SPCA) or similar society, with no private enforcement right."[62] Few public prosecutorial agencies will view animal welfare as a high priority, and civil enforcement faces standing problems.[63] A New York court eloquently summarized this situation:

> The reluctance or inability on the part of the defendant

---

[56] ANIMAL WELFARE INSTITUTE, supra note 13 at 10.
[57] WASH. REV. CODE ANN. § 16.52.070 (West 1994).
[58] P.R. LAWS ANN. tit 5, § 1652 (1984).
[59] States that specifically exclude poultry are Louisiana and South Carolina, see Appendix. See also Wise, supra note 26, at 206.
[60] DANIEL S. MORETTI, *ANIMAL RIGHTS AND THE LAW* 6-7 (1984)
[61] Id. at 6; Friend, supra note 23 at 215-220. While it is difficult to research such a topic as any case research will not reflect the amount of guilty pleas, a search provides some insight. A Lexis search in 1996 for cruelty to farm animals in all states since 1970 found only six cases.
[62] Wise, supra note 26, at 206.
[63] See Friend, supra note 23, at 215-218.

ASPCA as set forth above, raises serious questions, vis-a-vis the effectiveness of our present procedure for dealing with allegations of cruelty to farm animals on the large scale. However, refinement or amendment of this procedure is in the province of the legislature rather than this court...It's ironic that the only voices unheard in this entire proceeding are those of innocent, defenseless animals.[64]

Convictions are infrequent and generally limited to minimal fines. For example, Montana has a fine of $500,[65] while New Jersey imposes a fine of $250 for general cruelty to animals.[66] With little enforcement and small penalties, many individuals can only view such laws as irrelevant.

It is also extremely difficult to ascertain what occurs on the average farm, because a farm is private property. Police and law enforcement officers associated with SPCAs and humane societies must demonstrate probable cause to obtain a warrant to search private property for evidence of abuse. Unless the agency is informed by someone "on the inside," it is extremely difficult for information to be discovered, and evidence obtained by a humane officer without a valid warrant will be suppressed.[67] For example, recent evidence concerning "downers" abuse (animals crippled before or during transportation and then dragged) only surfaced when a private individual, Becky Sandstedt, conducted 18 months of hidden video-taping of handling abuses. The video-tapes were publicized by Farm Sanctuary, and gained national attention when the tapes were played on the TV news show Expose.[68]

Most importantly, there is a significant trend within states to remove legal protection from animals raised for food or food production altogether; if a farming practice is viewed by the agriculture industry as "accepted," "common," "customary" or "normal," the anticruelty statute will not be applied. The limited protections outlined above, which with greater enforcement could be somewhat effective in providing protection to animals raised for food or food production from institutionalized cruelty, are of no use if the treatment of such animals is specifically exempted from the coverage of the state statute, and thus condoned.

---

[64]County of Albany v. ASPCA, 447 N.Y.S.2d 662, 664, 112 Misc.2d 829, 832 (1982).
[65]MONT. CODE. ANN. § 45-8-211(2) (1993).
[66]N.J. STAT. ANN. § 4:22-26 (1998).
[67]See e.g., State v. Osborn, 63 409 N.E.2d 1077 (Ohio. 1980).
[68]Downer Cows: Have Things Changed?, Vol. 29, BEEF, No. 6, February 1993; see also Friend, supra note 23.

This pattern of amendments begs the obvious question of what such customary farming practices are. If such practices are cruel and cause pain and unnecessary suffering, state legislators are limiting the application of anticruelty laws to allow cruelty to occur. Moreover, even if today's customary farming practices are determined not to be cruel, the legislation of such exemptions creates an arena whereby farming practices can be developed, without fear of sanction, regardless of how cruel they may be.

III. "ACCEPTED," "COMMON," "CUSTOMARY" OR "NORMAL" FARMING PRACTICES

The question is not, can they reason? Nor, can they talk? But can they suffer? Why should the law refuse its protection to any sensitive being? The time will come when humanity will extend its mantle over everything which breathes...[69]

A. On the Farm

Farming practices in the United States dictate the fate of the majority of animals that come into contact with humans. Almost eight billion animals are killed annually in the United States for food,[70] as compared with 220 million from hunting (mostly birds and small animals), 5.4 million cats and dogs in pounds (although this number is somewhat disputed and may be higher), over 20 million through research and testing,[71] 18 million by dissection in classrooms and 4.5 million for fur

---

[69] Jeremy Bentham, Principles of Morals and Legislation, in THE EXTENDED CIRCLE, supra note 4, at 23.

[70] According to the statistics provided by the USDA for cattle, calves, hogs, sheep, lambs and poultry in 1996: 7.7 billion chickens, turkeys and ducks, 36.6 million adult cattle, 1.8 million calves, 4.2 million sheep and lambs, 92.4 million pigs. See Agricultural Statistics, United States Department of Agriculture (1998). These numbers do not include animals raised for food production, e.g., eggs and milk; for example, in 1997 there were approximately 300 million laying hens and 9 million dairy cows in the United States. Id.

[71] According to the Animal Welfare Act Reports for 1993: 49,561 primates, 106,191 dogs, 33,991 cats. Additionally, it is estimated that 392,000 guinea pigs, 318,000 hamsters, 426,000 rabbits, 365,000 farm animals, and 678,000 "others" are killed annually for research and testing, not including mice and rats which account for approximately 90% of the total. See Animal Welfare Enforcement Fiscal Year 1993, Report of the Secretary of Agriculture to the President of the Senate and the Speaker of the House of Representatives.

garments (of which two million are ranched and 2.5 million are trapped).[72] Approximately 20 million chickens and some 90,000 cows and calves are slaughtered every 24 hours in the United States.[73] The great majority of animals used for food or food production are raised using intensive husbandry practices. When discussing the treatment of such a large number of animals it is hard not to write either in a droning monotone or somewhat sensationally, but a brief analysis of a few customary farming practices is necessary to understand what a simple legal exemption actually achieves in practice. It is not simply eight billion animals a year; but it is one, and one, and one, amounting to the large scale mistreatment of individual animals.[74]

For example, pigs are castrated and have their tails removed without an anesthetic. Moreover, gestating (pregnant) sows and farrowing (birthing) sows are housed in stalls where they are unable to turn around ("gestation crates" or "farrowing crates"). Such intensive farming practices result in health problems, including lameness or high death losses, aggravated by indiscriminate genetic selection for production traits such as rapid growth. Genetic problems are increasing; some pigs are so excitable that quiet humane handling at the slaughter plant is very difficult.

Agribusiness subjects cattle of all ages to inhumane practices. For example, day-old baby calves are transported from the dairy farm before they are able to walk, resulting in calves being thrown, dragged, or trampled. This practice is becoming increasingly accepted at dairies in some parts of the country. Furthermore, cattle farmers often drag downed, crippled cows and will sell cows for slaughter when they are physically unfit to travel. Some communities consider this an accepted practice, but most good producers condemn the abuse of downers. Most

---

[72]Telephone interview with Merritt Clifton, Editor, *Animal People*, (1995).

[73]What Humans Owe to Animals, *The Economist*, August 19, 1995, at 11; see also, supra, note 70.

[74]The examples of farm practices in the following portion of the text were obtained from the following sources: J.R. GILLESPIE, MODERN LIVESTOCK AND POULTRY PRODUCTION (2d ed. 1983); B.P. SMITH, LARGE ANIMAL INTERNAL MEDICINE (1990); ANIMAL WELFARE INSTITUTE, supra note 13; ANIMAL WELFARE INSTITUTE, FACTORY FARMING: THE EXPERIMENT THAT FAILED (1987); J. MASON AND P. SINGER, ANIMAL FACTORIES (1990); J.B. MASON, INTENSIVE HUSBANDRY SYSTEMS, ANIMAL FOOD PRODUCTS AND HUMAN HEALTH (1991); J. Mench and A. Van Tienchoven, FARM ANIMAL WELFARE, AMERICAN SCIENTIST, Nov/Dec. 1986, at 598; Grandin, supra note 131; and agricultural journals such as BEEF, NATIONAL HOG FARMER, and FEEDSTUFFS.

downer cows are emaciated or in poor physical condition before they leave the farm. Veal calves are housed in stalls where they are unable to turn around. The calves are fed a liquid diet that does not allow the normal function of the calf's rumen. In addition, cattle are dehorned, castrated and hot-iron branded without anesthetic.

Poultry are also the victims of cruel husbandry practices, such as the

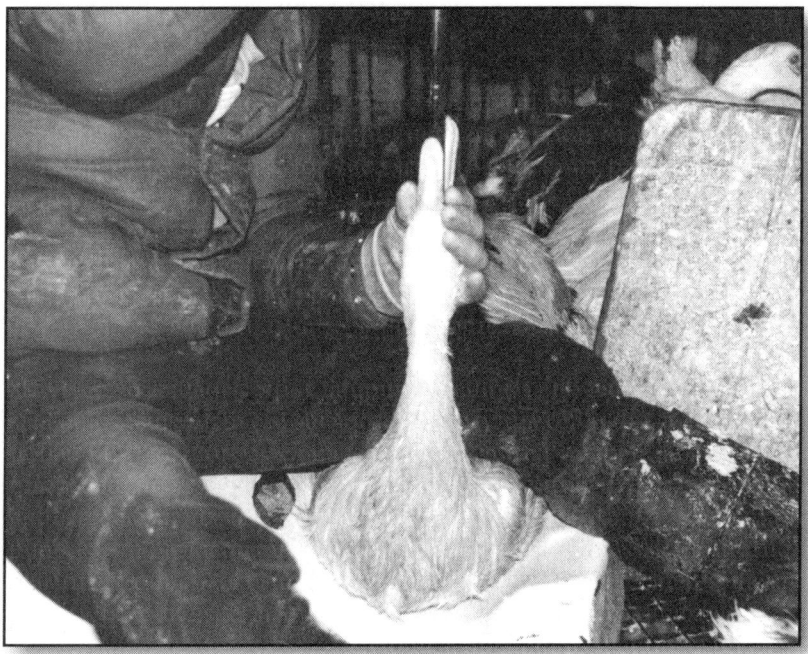

removal of chicken's beaks with hot cauterizing blades. Additionally, the starvation of laying hens to make them enter the next laying cycle is a common practice. This is termed "forced molting." Egg layers are housed without access to a nest box in a manner that does not allow the birds a full range of motion (the "battery cage"). Another common practice is the disposal of male chicks or live unhatched eggs by suffocation. Agribusiness does not restrict its cruel practices to chickens. For example, geese are force-fed for the foie gras trade by pump-feeding food down the birds' throats.

Further evidence that common farming practices are cruel can be found in the recent English "McLibel" holding. Following a lengthy and detailed review of common farming practices in the United Kingdom and the United States, Mr. Justice Bell found the following common

Above: A goose is manually force-fed in order to swell its liver for paté.

farming practices proven to be cruel: the battery cage; the restriction of movement in the last few days of the lives of broiler chickens; the restriction of sows in gestation crates; the cutting of the throats of fully conscious chickens; calcium deficits resulting in brittle bones in battery hens; the restriction of feed to broiler chickens bred for appetite; leg problems in broiler chickens bred for weight; the rough handling of broiler chickens and pre-stun electric shock suffered by broiler chickens on their way to slaughter; and the gassing of chicks by carbon dioxide.[75]

B. Transport and Slaughter

Transportation and slaughter also cause suffering to animals raised for food or food production. The following is a brief list of examples: horses are transported in double-decker cattle trucks with a ceiling so low that they injure their heads and backs; animals are transported on long journeys without water or rest stops; animals are bred in a manner that produces genetic factors which increase death losses; conscious animals are shackled and hoisted by one back leg prior to ritual slaughter. Abuses in the slaughter process have been extensively documented. Finally, genetic selection of animals for rapid weight gain and other traits results in very excitable pigs and cattle who are extremely difficult to move in a quiet manner at the slaughter plant. Animals will sometimes refuse to move quietly through state of the art facilities that work well for normal animals. This results in pile ups and abuses by frustrated handlers.[76]

---

[75]McLIBEL VERDICT, Sect. 8; The "McLibel" case began in 1991 when McDonald's sued five individuals in England for defamation in relation to statements made in a pamphlet produced by an organization called London Greenpeace. The pamphlet, entitled, "What's Wrong With McDonald's? Everything They Don't Want You to Know," accused McDonald's of a variety of unpleasant practices, including, among other things, being responsible for the destruction of the rainforest, heart disease, cancer, food poisoning, and cruelty to animals. Two of the individuals (Helen Steel and Dave Morris) refused to settle with McDonald's and represented themselves *pro se* during the course of a three-year trial, and through appeals which still continue. Despite numerous disadvantages facing the defendants, McDonald's lost significant portions of the case at the trial and appeal level. Most importantly for the purposes of this booklet, the court held that the defendants had told the truth when they stated that McDonald's was responsible for the large-scale mistreatment of certain animals raised for its food products. For further information about McLibel and its impact on animal legal issues, see David J. Wolfson, McLibel, 5 ANIMAL LAW (1999), and for general information see the excellent website www.mcspotlight.org.
[76]Supra, note 74; Eisnitz, supra, note 14.

IV. STATE ANTICRUELTY STATUTES THAT PROHIBIT THEIR APPLICATION TO "ACCEPTED," "COMMON," "CUSTOMARY" OR "NORMAL" FARMING PRACTICES

As of today, 30 states' anticruelty statutes specifically exempt all or some customary farming practices, such as those described in the preceding section.[77] In some cases, the transport of animals raised for food is also exempted.[78] Additionally, the Texas anticruelty statute contains a bizarre provision whereby it is a crime to kill, injure, or administer poison "to an animal, other than cattle, horses, sheep, swine, or goats, belonging to another without legal authority or the owner's effective consent."[79]

Twenty-five of the 30 states referred to above prohibit the application of their anticruelty statutes to all customary farming practices.[80] Moreover, of these 30 states, 18 amended their statutes in the last ten years to place agribusiness beyond the statutes' reach.[81] In the last four years alone, Connecticut, Idaho, Iowa, Michigan, New Jersey, North Carolina, and Wyoming enacted amendments to their anticruelty statutes to exclude animals raised for food or food production.[82] Clearly, a definite trend exists.[83]

---

[77] See Appendix.

[78] It is arguable that the definition of "accepted," "common," "customary" or "normal" farming practices would include "accepted," "common," "customary" or "normal" methods of transportation of animals raised for food or food production.

[79] TEX. PENAL CODE ANN. § 42.09(5) (West 1996). New Mexico has also enacted an illegal confinement statute whereby it is a criminal offense to intentionally separate offspring of a livestock animal from its mother, "provided that, when milk cows, which are actually used to furnish milk for household or dairy purposes, have calves, that are unbranded, such young animals may be separated from their mother and enclosed." N. M. STAT. ANN § 30-18-5(C) (Michie 1995).

[80] Those states are: Arizona, Colorado, Connecticut, Idaho, Illinois, Indiana, Iowa, Kansas, Maryland, Michigan, Missouri, Montana, Nebraska, Nevada, New Jersey, North Carolina, Oregon, Pennsylvania, South Carolina, South Dakota, Tennessee, Utah, Washington, West Virginia and Wyoming.

[81] Those states are Colorado, Connecticut, Idaho, Indiana, Iowa, Michigan, Montana, Nebraska, Nevada, New Jersey, North Carolina, South Carolina, South Dakota, Tennessee, Utah, Vermont, West Virginia and Wyoming.

[82] CONN. GEN. STAT. ANN. § 53-247(9)(1998); IDAHO CODE § 25-3514(5)(9) (1997); IOWA CODE § 717.2 (1997); MICH. STAT. ANN. § 28.245 (2) (1997); N.J. STAT. ANN. § 4: 22-16.1 (1998); N.C. GEN. STAT. § 14-360(c)(2) (1998); WYO. STAT. § 6-3-203 (f) (u) (1997).

[83] This article will refer to amendments that "exempt" or "exclude" animals raised for food. It is worth noting, as a legal matter, that some of the statutes listed above are unclear as to whether they simply exempt or exclude such animals or, in the

Following are several examples of state law exemptions for farming practices: "nothing shall affect the accepted animal husbandry practices utilized by any person in the care of livestock animals;"[84] "nothing in this act affects normal good husbandry practices utilized by any person in the production of food;"[85] exemption for "commonly accepted agricultural and livestock practices on livestock;"[86] and the act does not "prohibit or

alternative, provide an affirmative defense for a defendant who can prove that the alleged cruel act is in fact an "accepted," "common," "customary" or "normal" farming practice; for example, the New Jersey statute states that "there shall exist a presumption that the raising, keeping, care, treatment, marketing, and sale of domestic livestock in accordance with the standards developed [by the State Board of Agriculture and the Department of Agriculture, in consultation with the New Jersey Agricultural Experiment Station] shall not constitute a violation of any provision of this title involving alleged cruelty to, or inhumane care or treatment of, domestic livestock." N.J. STAT. AM § 4:22-16.1 (1998). The practical consequence, however, is the same: the farming practice will not be successfully prosecuted under the relevant anticruelty statute.

[84]COLO. REV. STAT. § 18-9-201.5 (1997).
[85]ILL. ANN. STAT. 410 ILCS 70/13 (1998).
[86]MONT. CODE ANN. § 45-8-211(4)(a) (1997).

Above: Turkeys are raised in intensive confinement warehouses.

interfere with established methods of animal husbandry including the raising, handling, feeding, housing, and transporting, of livestock or farm animals."[87]

A notable example of these exemptions are the amendments recently enacted in Idaho and Iowa. The Idaho amendment provides that the anticruelty statute shall not be construed to interfere with normal or accepted practices of animal husbandry[88] and includes an additional section that states:

> any other...activities normally or commonly considered acceptable...[are exempted]...all activities described in this section are not construed to be cruel nor shall they be defined as cruelty to animals, nor shall any person engaged in the practices, procedures, or activities be charged with cruelty to animals.[89]

Similarly, Iowa recently amended its anticruelty statute. The new statutory provisions provide a graphic example of how such amendments strip away pre-existing legal protection from animals raised for food or food production. Prior to the 1994 amendment, Iowa had two general anticruelty sections within a chapter entitled "Injury to Animals."[90] The 1994 amendment created two new chapters: "Injury to Livestock" and "Injury to Animals Other than Livestock."[91] Thus, on first impression, the amendment appears to provide a greater degree of protection to livestock than existed prior to its enactment. Such a conclusion is incorrect.

Specifically, the chapter entitled "Injury to Livestock" contains two subsections, the first of which only applies to livestock owned by another person. By contrast, the third subsection applies to all livestock, but only if the livestock is provided with care inconsistent with "customary animal husbandry," or if livestock is injured or destroyed "by any means which causes pain or suffering in a manner inconsistent with customary animal husbandry practices."[92] While the second subsection does provide that livestock must not be deprived of "necessary sustenance," the first subsection defines "sustenance" as "food, water, or a nutritional formulation customarily used in the production of livestock."[93]

---

[87] NEV. REV. STAT. ANN. § 574.200.6 (1997).
[88] IDAHO CODE § 25-3514(5) (1997).
[89] Id. § 25-3514(9) (1997).
[90] IOWA CODE §§ 717.1, 717.2 (1997)
[91] IOWA CODE §§ 717, 717 (B) (1997).
[92] Id. § 717.2(1)(a),(c) (1997).
[93] Id. § 717.2(1)(b), § 717.1(6) (1997).

The remaining two sections enacted pursuant to the 1994 amendment are in the chapter entitled "Injury to Animals Other than Livestock." The definition of animal for the purposes of these two sections specifically exempts "livestock."[94] Thus, a total evaluation of the amendment reveals that "customary farming practices" are exempted from the reach of the anticruelty statute. Also, "livestock," comprising all "bovine, caprine, equine, ovine or porcine species or poultry," is no longer included within the definition of "animal" in the general section relating to cruelty to animals.[95]

In 1995, New Jersey also amended its anticruelty statute. The new statutory language states that the "raising, keeping, care, treatment, marketing and sale of domestic livestock" shall be legally presumed to not be cruel if livestock are kept in accordance with standards developed and adopted by the "State Board of Agriculture and the Department of Agriculture in consultation with the New Jersey Agricultural Experiment Station."[96] The statute mandated that these standards were to be adopted, pursuant to the New Jersey Administrative Procedure Act, within six months of the enactment of the amendment. As of April 1999, however, New Jersey had failed to develop or adopt such guidelines.

The most recent state to prohibit the application of its anticruelty statute to customary farming practices is North Carolina, which amended its anticruelty statute in 1998. The amendment exempts any "lawful activities conducted for...purposes of production of livestock or poultry," but does not define "lawful activities" or provide any guidance as to what a "lawful" activity is.[97] It is unclear how a prosecutor would determine whether to initiate a criminal prosecution against a customary farming practice.

By contrast, certain states exempt only specific practices instead of all customary farming practices. Wisconsin does not require animals raised for food or food production to be provided shelter other than as provided by normally accepted husbandry practices in each particular county.[98] Ohio exempts such animals from requirements for wholesome exercise and a change of air.[99] Virginia's anticruelty statute states that it shall not be construed to prohibit the dehorning of cattle.[100] Vermont

---

[94]Id. § 717B.1(1)(a) (1997).
[95]Id. §§ 717.1(2), 717B.1 (1)(a) (1997).
[96]N.J. STAT. ANN. § 4-22-16.1(1998).
[97]N.C. GEN. STAT. § 14-360(c)(2)(1998).
[98]WIS. STAT. ANN. § 951.14 (1997).
[99]OHIO REV. CODE ANN. § 959.13 (A)(4) (Anderson 1998).
[100]VA. CODE ANN. § 3.1-796.122(C) (1997).

exempts animals raised for food or food production from the section in its anticruelty statute that deems it illegal to "tie, tether and restrain" an animal in a manner that is inhumane or detrimental to its welfare.[101] Additionally, Louisiana and South Carolina exclude poultry from the protection of their anticruelty statutes, thus removing state legal protection for animals which represent at least 95% of the animals killed for food every year in the United States.[102]

The effect of this trend of amendments cannot be overemphasized. The trend indicates a nationwide perception that it was necessary to amend anticruelty statutes to avoid their possible application to animals raised for food or food production. Amendments specifically exempting customary husbandry practices indicate that, but for the exemption, such practices would be determined to be cruel. Exemptions for specific practices, such as tethering, exercise or shelter clearly demonstrate this attitude. Who decides what is considered a customary practice? Seemingly, the definition of an inhumane practice is determined by the average farmer. In Wisconsin[103] the county determines the customary practice, in New Jersey, the "State Board of Agriculture and the Department of Agriculture in consultation with the New Jersey Experiment Station"[104] and Tennessee provides for determination by a "college of agriculture or veterinary medicine."[105] Legislatures have endowed the agribusiness community with complete authority to define what is, and is not, cruelty to the animals in their care. Particularly striking is the recently enacted Idaho statute, which not only states that the anticruelty statute shall not be construed as interfering with accepted practices of animal husbandry or any "other normally or commonly considered acceptable" practice, but has also placed all powers of enforcement for the anticruelty statute with the Department of Agriculture.[106]

A further example of the power and intent of the farming community can be found in a 1997 amendment to the Tennessee anticruelty statute, which removed the power to investigate allegations of cruelty to farm animals from agents of a society to prevent cruelty to animals. The amendment, however, allowed such societies to retain their ability to investigate cruelty in relation to "non-livestock animals."[107] In addition,

---

[101] VT. STAT. ANN. tit. 13, § 352(a)(3) (1998).
[102] According to the USDA's statistics, 177 million broiler chickens were slaughtered in South Carolina in 1996. Supra, note 71.
[103] WIS. STAT. ANN §951.14 (1997).
[104] N.J. STAT. ANN § 4:22-16.1 (1998).
[105] TENN. CODE ANN § 39-14-202(c)(1) (1997).
[106] IDAHO CODE § 25-3501 (1997).
[107] TENN. CODE. ANN. § 39-14-210(a) (1997).

the amendment stated that no individual could enter the property of another, arrest or interfere with "usual and customary agricultural or veterinary practices," or confiscate livestock or take any other action in response to an allegation of cruelty to livestock unless, prior to or at the same time as taking any such action, the livestock in question was "examined by the county agricultural extension agent of such county, a graduate of an accredited college of veterinary medicine specializing in livestock practice or a graduate from an accredited college of agriculture with a specialty in livestock."[108] No further action or investigation may be taken unless such an individual has probable cause to believe a violation of the statute occurs.[109]

This amendment is remarkable, and disturbing, in several respects. First, since a society for the prevention of cruelty to animals cannot investigate cruelty to farm animals, who is meant to be enforcing the statute? Presumably, the police, who are overburdened and have no particular interest in preventing animal abuse, or the Department of Agriculture, whose primary purpose is not animal well-being. The most concerned and active enforcer of the anticruelty statute, a society for the prevention of cruelty to animals, is only deemed responsible enough to protect companion animals. Second, even a graduate of a veterinary or agriculture college is not presumed to be able to recognize animal abuse or cruelty unless he/she "specialized" in livestock practice. The most likely determiner of cruelty will be the "county agriculture extension agent" who has a disincentive to enforce the statute because of his/her relations with the farming community in the county. Third, the statute seems to invest the "extension agent, veterinary college graduate specializing in livestock practice or livestock specialist" with the sole authority to determine whether "probable cause" exists to pursue an investigation, as opposed to the police or local district attorney.[110] All of these protective mechanisms occur in a state that already exempts usual and customary practices which are accepted by colleges of agriculture or veterinary medicine from the anti-cruelty statute. It is hard to imagine

---

[108]Id. § 39-14-211 (1997).
[109]Id.
[110]Id. The statute does provide that if a person authorized to make an inspection does not do so within twenty-four hours of receipt of a complaint, a licensed veterinarian (who does not have a specialty in livestock) may make such inspection. It is unclear how the veterinarian would become aware of the need for such inspection. Furthermore, the power to investigate is only permissive, i.e., the statute does not require such a veterinarian to make an investigation if a specialty veterinarian does not investigate within 24 hours of a complaint.

how the farming industry could more successfully control the definition of what is cruelty to farm animals and, at the same time, more effectively build a barricade to prevent anyone but the farming industry discovering what actually occurs on the farm. The role of a prosecutor or judge, the normal participants in the enforcement of a criminal statute, is non-existent.

A similar bias can be found in Virginia, where a prosecution for cruelty to "non-agricultural animals" can be initiated within five years following the cruel act, but must be initiated within one year in the case of cruelty to "agricultural animals."[111]

The most remarkable aspect of the statutes discussed in this section is that seemingly any practice considered customary cannot be successfully prosecuted on the basis of cruelty. As Idaho's law states, normal or commonly accepted animal husbandry and other practices "shall not be construed to be cruel nor shall they be defined as cruelty to animals, nor shall any person engaged in these practices, procedures or activities be

---

[111] VA. CODE ANN. § 3.1-796.122 (1997).

Above: A calf collapses at an auction yard.

charged with cruelty."[112]

For example, in a recent Pennsylvania case, a defendant accused of starving his horses argued that the practice of denying nutrition to horses that were no longer wanted and were to be sold for meat was a "normal agricultural operation," not a criminal act.[113] The defendant elicited testimony from witnesses who stated it was a normal practice "to neglect...horses for sale...for meat."[114] Such horses, the defendant argued, are commonly denied veterinary care and sufficient nutrition, and they are placed in so-called "killer pens." Witnesses also stated that "various practices in the farming industry...might be considered cruel except for the fact that they are practices within the industry" (like the raising of veal calves and chickens).[115]

The court convicted the defendant of cruelty. The defendant failed to establish sufficient testimony as to the pervasiveness of the practice, and no testimony "indicat[ed] that in fact they were in the business of raising horses to be sold for dog food or that they had formed the definite intention of sending the horses in question to 'killer pens' for that purpose."[116]

This case highlights two essential issues related to the exclusion of customary farming practices from anticruelty statutes. First, if the defendant had successfully shown with additional testimony that the practice was a normal practice, the anticruelty statute would not have applied to the act of starving his horses, and the court would not have found him criminally liable. The defendant's problem was not that he starved his horses, but that he could not prove that enough people were doing the same thing. Clearly, if everyone does it, anything is possible under the new statutes. Second, if the defendant had proved he intended to be cruel for an economic reason, the court would have been less likely to convict him. In part, it was his lack of both intent and motive for profit that resulted in the criminal conviction.

Ultimately, many of the examples of customary farming practices described in the preceding section constitute cruelty to animals raised for food or food production, and state anticruelty laws that cover animals raised for food or food production are not applied to these practices. It is also clear that a large number of legislatures in the United States have created legal exemptions to allow such cruelty to continue. In effect, state

---

[112]IDAHO CODE § 25-3501, § 25-3514(5)(9) (1997).
[113]Commonwealth v. Barnes, 629 A.2d 123 (Penn. 1993).
[114]Id. at 130.
[115]Id. at 132.
[116]Id.

legislatures have recognized that without amending anticruelty laws, many of the practices described in the preceding sections could be criminal offenses.

Many European legislatures have also recognized that many of the customary farming practices described above are cruel. However, instead of altering the law so as to exempt the practices from legal protection and criminal prosecution, many Western European countries have outlawed the same practices.

V. THE SITUATION IN EUROPE

If you have men who will exclude any of God's creatures from the shelter of compassion and pity, you will have men who will deal likewise with their fellow men.[117]

A. Origins of European Anticruelty Laws

In the Europe of the Middle Ages, man viewed himself as having God-given dominion over the world, although this was perceived as more of a feudal stewardship rather than one of natural domination.[118] In this world, animals and humans had an interactive relationship; the louse "existed to prompt humans to be clean, and the irksome horse-fly to stimulate man's ingenuity."[119] Animals were even the subject of legal trials in continental Europe. Examples can be found of a "sow being mutilated and hanged after it had killed a child, and leeches being excommunicated for killing fish in Lake Geneva."[120] As a legal commentator has stated, "by integrating animals within a human scheme of justice, these trials allowed the community to affirm a rational order and assign a role for animals within the hierarchy of creation."[121]

With the Enlightenment and the growth of towns and commerce

---

[117]St. Francis of Assisi, quoted in St. Bonaventura, the Life in THE EXTENDED CIRCLE, supra note 4, at 325.
[118]Man's Mirror. (History of Animal Rights),The Economist, Nov. 16, 1991, at 22.
[119]Id. at 23.
[120]Id.
[121]Paul Schiff Berman, Rats, Pigs and Statutes on Trial: The Creation of Cultural Narratives in the Prosecution of Animals and Inanimate Objects, 69 N.Y.U.L. REV. 288, 291 (1994).

there arose what has been described as a "contractual notion of ethics: do as you would be done by."[122] The contract relationship did not extend to animals, but the notion of rationality began to enter into the animal-human relationship. Habits changed with the increase in household pets during the eighteenth century. Rousseau and Voltaire condemned man's treatment of animals as well as man's treatment of

man. In 1824, the animal welfare movement came of age when William Wilberforce and Sir Thomas Fowell Buxton, two leaders of the movement to abolish the slave trade, both assisted in founding the Royal Society for the Prevention of Cruelty to Animals.[123] The first anticruelty law in Britain was passed in 1822 and condemned baiting and beating.[124]

Interestingly, early twentieth century English case law provides

---

[122]Man's Mirror. (History of Animal Rights), supra note 117, at 24.
[123]Id.
[124]Id. See also, Favre & Tsang, supra note 19.

Above: Pigs who died in transit are left behind a truck at a slaughterhouse.

precedent for the proposition that cruel customary farming practices should be prohibited. In a 1913 case, Waters v. Braithwaite,[125] it was argued that the failure to milk a cow for twelve hours, causing distended udders to demonstrate that the cow was a good milker, was not a cruel act because the practice was "customary" and performed for a commercial purpose. In reply, Justice Darling found the practice to be in violation of the Protection of Animals Act, stating that

> [i]t was not denied that [the practice] caused great pain; no one alleged that it produced any benefit to the cow...[t]he only benefit there was might be that of the owner.... If the custom of doing this did exist, it was time that it ceased, and people must find some other means of judging whether a cow was a good milker.[126]

B. The Situation Today

Recent European concern over the conditions of intensive farming of animals began to arise shortly after the publication of a book by Ruth Harrison entitled *Animal Machines*, in 1964.[127] The book prompted the British government to appoint a committee "to examine the conditions in which livestock are kept under systems of intensive husbandry and to advise whether standards ought to be set in the interests of welfare, and if so what they should be." This Committee, the Brambell Committee, set forth the "Five Freedoms" of movement:

> In principal we disapprove of a degree of confinement of an animal which necessarily frustrates most of the major activities which make up its natural behavior.... An animal should at least have sufficient freedom of movement to be able without difficulty to turn around, groom itself, get up, lie down, stretch its limbs.[128]

While none of these recommendations were given the force of law, their effect was significant.

---

[125]30 T.L.R. 107, 108 (K.B. 1913).
[126]Id. at 108.
[127]Wise, supra note 26, at 211.
[128]Id. at 212.

Specifically, in 1987, the Parliament of the United Kingdom banned the veal crate and the anemic diet for veal calves. The Welfare of Calves Regulation 1987 Act reads:

> No person shall keep, or knowingly cause or permit to be kept, a single calf in a pen or stall on any agricultural land unless the following requirements are complied with:
>
> (a) the width of the pen or stall is not less than the height of the calf at the withers; (b) the calf is free to turn round without difficulty; (c) the calf is fed a daily diet containing sufficient iron to maintain it in full health and vigor; (d) if the calf is more than 14 days old, it has access each day to food containing sufficient digestible fiber so as to not impair the development of its rumen.[129]

Similarly, the Pig Husbandry Law was enacted in 1991, making it illegal to rear sows in cramped stalls after 1999, when the law comes into effect.[130] The Welfare of Livestock Regulations of 1982 also prohibits short tail docking of sheep, hot branding of cattle, penis amputation, tongue amputation in calves, and tail docking of cattle; the Docking of Pigs Regulations of 1974 prohibits the docking of the tail of pigs more than seven days old without anesthetic.[131] Additionally, England forbids the sale of day-old calves.[132]

In Western Europe, there has been further legal precedent for finding customary farming practices unacceptable. The German Animal Protection Act of 1972, revised in 1986, prohibits force feeding an animal except for health reasons.[133] Moreover, a West German appellate court "ruled that keeping laying hens in battery cages violated the German Animal Protection Act of 1972, as the practice failed to take the natural behavior of hens into account."[134] In 1979, the Frankfurt Court of Appeals ruled that the use of battery cages, as used in West Germany,

---

[129]Welfare of Calves Regulations No. 2021 (U.K. 1987), cited in ANIMAL WELFARE INSTITUTE, supra note 13, at 304.
[130]The Meat of the Matter, THE ECONOMIST, Jan. 21, 1995, at 58.
[131]Glen H. Schmidt & Beverly A. Schmidt, Animal Welfare Legislation in Northern Europe 25 (1995) (unpublished article, on file with author). SIMON BROOMAN & DR. DEBBIE LEGGE, LAW RELATING TO ANIMALS, 199 (1997).
[132]Temple Grandin, Farm Animal Welfare during Handling, Transport and Slaughter, 204 J. AMER. VET. MED. ASS'N 372, 373 (1994).
[133]Schmidt & Schmidt, supra note 130, at 34.
[134]Wise, supra note 26, at 212. This act has been replaced by the Animal Protection Act of 1986 which came into effect on January 1, 1987.

constitutes cruelty within German Federal Law and was punishable under Section 17.2b of the German Animal Protection Act.[135]

In Switzerland, the Animal Protection Act, which became operative in July 1981, banned all battery cages by the end of 1991. The law requires housing systems for laying hens to provide sheltered, darkened nesting boxes and perches or slatted grids for all hens and allows a minimum of eight hundred square centimeters per bird, thus effectively prohibiting the keeping of laying hens in cages. The method of choice in Switzerland is now the aviary, "conceived in accordance with the natural behavior of fowl and based on installations and equipment such as nest boxes and scratching areas, or perches that enable birds to follow patterns of behavior specific to their species."[136]

The Swiss Animal Protection Regulations of May 27, 1981 also provide that animals shall not be permanently tethered and "stalls, boxes and tethering systems shall be so designed that animals can lie-down, rest and rise to their feet in the way normal for their species."[137] Furthermore, calves must receive sufficient iron in their feed, and chickens selected for killing shall not be piled on top of one another while still alive.[138] Moreover, animals shall not be transported unless they can be expected to withstand the journey without harm.[139] Additionally, producers in Denmark have to pay a rendering truck to remove downers, which are not allowed at the slaughter plant.[140]

The European Convention for the Protection of Animals Kept for Farming Purposes, adopted by the Council of Europe and recently ratified by the European Union, "requires that animals be housed and provided with food, water, and care in a manner which is appropriate to

---

[135]Id. In 1979, the original claim raised in the district court was rejected only to be overturned at the appellate level. In 1985, the district court rejected another charge of cruelty.
[136]1981 Swiss Ban on Battery Cages: A Success Story for Hens and Farmers, 44 ANML. WELFARE INST. Q. , No. 1, at 10. The information reported in the above publication is based on a report: Laying Hens: 12 years of experience with new husbandry systems in Switzerland, Swiss Society for the Protection of Animals (on file with author).
[137]Council of Europe—Information Document; Swiss Animal Protection Regulations, May 27, 1991, art. 6 (on file with author); Wise, supra note 26, at 212.
[138]Id. art. 16, 20, 26.
[139]Id. art. 54.
[140]Grandin, supra note 131, at 372-373. In New Zealand, a downed animal cannot be sent to a slaughter plant until it is inspected on the farm by a veterinarian; in Australia, downed cattle that arrive at the slaughter plant at night are often euthanized and sent to rendering; and in Canada, large slaughter plants have stopped accepting downers.

their physiological and ethnological needs, taking the species into consideration."[141] Article 4 places limits on the restriction of freedom of movement which causes the animal unnecessary suffering or injury, also taking each species into account.[142] On April 28, 1999, the European Union extended its ban on United States beef to encompass a total ban on all United States beef imported into the European Union because of health fears over the use of hormones in the United States; these hormones are used to stimulate growth in cattle. In addition, the European Union has enacted a moratorium in the European Union on the use of a genetically engineered hormone, bovine somatropin, that increases the output of milk in dairy cattle and hastens growth.[143]

On January 20, 1987, the European Parliament passed a report on animal welfare which would ban keeping veal calves in individual crates, phase out poultry battery cages within ten years, discontinue close confinement of pregnant sows, and ban routine tail-docking and castration of pigs. The European Commission has responded to the concerns in the report with a number of proposals. These proposals would take the form of a Regulation having direct effect and preempting national laws.[144]

It seems likely that the veal crate will be banned within the European community: Sweden, Denmark, Austria, Ireland, Finland, Belgium, and the Netherlands are in favor of abolishment. Additionally, the European Commission has condemned the use of individual veal crates. In a report, adopted on December 15, 1995, the European Commission stated that calves suffered severe stress when they were confined in crates and recommended the provision of adequate nutrients such as iron and roughage. The European Commission recommended that the crates be abolished by 2007.[145] The European Commission has also enacted rules stipulating hens in battery cages should have more space.[146] And, on June 15, 1999, agriculture ministers from the European Union agreed to end all battery egg production across the European Union from 2012. The system will be replaced by free-range farming, the housing of hens in

---

[141]Wise, supra note 26, at 212.
[142]Id. at 212-213; ANIMAL WELFARE INSTITUTE, supra note 13, at 292.
[143]Steve Lohr, Swedish Farm Animals Get a Bill of Rights, N. Y. TIMES, Oct. 25, 1988, at A1, A8; Stephen Castle, Europe Bans U.S. Beef Over Safety Fears, THE INDEPENDENT, Apr. 29, 1999, at 5.
[144]ANIMAL WELFARE INSTITUTE, supra note 13, at 286 (European Parliament Resolution on Animal Welfare Policy); Caroline Jackson, Europe and Animal Welfare, in ANIMAL WELFARE AND THE LAW 221-246 (1989).
[145]The Reuter European Community Report, December 19, 1995 (on file with author).
[146]Deborah Hargreaves, Commission Struggles to Hold Ring on Animal Welfare in EU, Financial Times, Jan. 12, 1994, at A3.

large, barn-like aviaries, or by cages with at least 750 square centimeters of space per chicken (compared with the current European norm of 450 square centimeters and 310 square centimeters in the United States), a nesting area with litter, a scratching pad to sharpen claws, and a perch. The Treaty of Rome, which established the European Community, was also recently amended to classify farm animals as sentient beings.[147]

Moreover, the European Union has recently enacted a Directive, which came into effect on January 1, 1997, enforcing an eight-hour maximum journey time for horses, cattle, pigs, sheep and goats being moved in standard haulage floats. Longer journeys can take place in vehicles with the following specifications: plenty of straw bedding, enough feed on board for the whole journey, direct access to the animals,

---

[147]Protocol on Improved Protection and Respect for the Welfare of Animals: "The High Contracting parties, desiring to ensure improved protection and respect for the welfare of animals as sentient beings, have agreed upon the following provision which shall be annexed to the Treaty establishing the European Community." Stephen Castle, EU Votes to End Battery Hen Farming in 12 Years, THE INDEPENDENT, June 16, 1999, at 5.

Above: A calf is chained in a crate for veal production.

forced ventilation available if necessary, moveable partitions to create separate compartments inside the vehicle, equipment to connect to a water supply during stops, and pig carriers must carry enough water for the whole journey.[148]

If these requirements are met, the following journey times will apply: calves and lambs still on milk and unweaned piglets can travel for a maximum of nine hours, one hour minimum rest for water and feed if necessary, then a further nine hours before 24 hours rest; pigs can travel 24 hours followed by 24 hours rest during which they must be unloaded, fed and provided water; adult cattle and sheep can travel for fourteen hours with one hour minimum rest for water and, if necessary, food, then 14 hours further travel is allowed with a final 24 hours rest.[149] Animals cannot be transported by sea or rail for more than eight hours, unless the facilities on the ship conform with those detailed in the previous paragraph.[150] Finally, member states will be able to introduce an eight-hour maximum journey time when transport to a slaughterhouse is entirely within a member state.[151]

Sweden deserves special attention. On May 27, 1988, the Swedish parliament passed a new animal protection law with three basic tenets. First, the law granted domestic animals the right to a favorable environment where their natural behavior is safeguarded. Second, protection for animals must improve, including protection from illness. Third, animal husbandry shall concentrate on keeping animals healthy and content.[152]

The law centered around the following regulations:

- All cattle are entitled to be put out to graze if over six months old.
- Poultry must be let out of battery cages.
- Sows may no longer to be tethered. They shall have sufficient room to move. Separate bedding, feeding and voiding places are to be provided. Breeding pigs should be given the opportunity to stay outdoors in the summer.

---

[148]Fordyce Maxwell, Reservations Over Animal Transport Deal, SCOTSMAN, June 23, 1995, at 26.
[149]Id.
[150]Id.
[151]Id.
[152]ANIMAL WELFARE INSTITUTE, supra note 13, at 305: Swedish Animal Protection, cited in Swedish Ministry of Agriculture Press Release (May 27, 1998).

- Cows and pigs must have access to straw and litter in stalls and boxes.
- Livestock buildings must be fitted with windows that let in light.
- Technology must be adapted to the animals and not the reverse. As a result it must be possible to test new technology from the animal safety and protection viewpoint before being put into practice.
- No drugs or hormones can be used on farm animals, except to treat disease.
- All slaughtering must be as humane as possible.
- In future, the government is empowered to forbid the use of genetic engineering and growth hormones which may mutate domestic animals.[153]

Most of the above requirements will be phased in over the next few years, although the implementation of the provisions making all chickens "free range" will be stretched over several years to lift some of the economic burden from farmers and provide time to build more spacious accommodation.[154]

In conclusion, a comparison shows that many common or normal farming practices in the United States are viewed as unacceptable in many European countries. Further evidence in support of this changing attitude in relation to cruel farming practices can be found in the determination of Mr. Justice Bell in the McLibel verdict (discussed on page 26). While it is unclear as to what extent all these laws are effectively enforced and implemented, many customary but cruel agricultural practices are now illegal or about to become illegal. This is in contrast to the United States, where the law is altered so as to avoid its legal and moral consequence. Thus if, as Mahatma Gandhi stated, "[t]he greatness of a nation and its moral progress can be judged by the way its animals are treated," the United States is being left behind.[155]

---

[153]Id.

[154]Impetus for passage of the Swedish law is usually attributed to an author of children's books, Astrid Lindgren. In 1985, Lindgren began writing satires about farm animal care in Sweden which were widely circulated in newspapers. Lohr, supra note 142, at A1.

[155]The Moral Basis of Vegetarianism, in THE EXTENDED CIRCLE, supra note 4, at 91.

VI. TOWARD AN ALTERNATIVE

> Until he extends the circle of his compassion to all living things, man will not himself find peace.[156]

A. The Cruel Reality

While the United States has a law in every state purporting to protect animals from cruelty, the amendments discussed above effectively withdraw such protection from the majority of domestic animals (animals raised for food or food production in those states). Concurrently, federal legislation focuses on protecting farming interests. The recently enacted Animal Enterprise Protection Act of 1992 is

---

[156]Dr. Albert Schweitzer, The Philosophy of Civilization, in THE EXTENDED CIRCLE, supra note 4, at 316.

Above: Battery cages intensely confine chickens for egg production.

designed to deter, prevent, and penalize crimes against farmers, ranchers, food processors, and agricultural researchers, with a penalty of imprisonment of up to a year.[157] The Attorney General and Secretary of Agriculture have the authority to "conduct a study on the extent and effects of domestic and international terrorism on enterprises using animals for food or fiber production, agriculture, research, or testing."[158]

The power of the farming industry in the United States must be recognized in order to fully understand the conflict that such a simple subject as the prevention of cruelty to animals raises and the difficulties facing any attempted reform. Even though less than two percent of the United States population is involved in producing the "raw materials" for the United States food supply, beef alone is a multi-billion dollar a year industry.[159]

Thus, farming is no longer a small family business. As Senator Metzenbaum declared while chairing a committee on disease in the poultry industry, "[t]he poultry industry is dominated by a few giant corporations, all of whom produce the same product with the same problems."[160] According to *Feedstuffs* (July 6, 1998), the top eight chicken processors' control over production increased from 25.3% in

---

[157]The statute sanctions anyone who "intentionally causes physical disruption to the functioning of an animal enterprise by intentionally, stealing, damaging, or causing the loss of any property (including animals or records)...and thereby causes economic damage exceeding $10,000 to that enterprise, or conspires to do so." 18 U.S.C. § 43 (1994). This federal legislation was supported by such groups as the National Livestock Producers Association, American Veal Association, National Board of Fur Farm Organizations, American Feed Industry Association, National Cattleman's Association, National Broiler Council, National Turkey Federation and the Pacific Egg and Poultry Organizations. Similar legislation has also been enacted in a number of states. See, e.g. ALA. CODE § 13A-11-150 (1990) (enacted in 1993); ARK. CODE ANN. § 5-62-201 (Michie 1993) (enacted in 1991); FLA. STAT. ANN § 878.40 (West 1994) (enacted in 1993); KY. REV. STAT. ANN. § 437.420 (Baldwin 1994) (enacted in 1990); OKLA. STAT. ANN. tit 21, § 1680.1 (West Supp. 1996) (enacted in 1991).
[158]Pub. L. No. 102-346 § 3(a).
[159]Schmidt & Schmidt, supra note 130 at 52.
[160]Poultry Safety: Consumers at Risk: Hearing on S.1324 Before the Senate Committee on Labor and Human Resources, 102nd Cong., 1st Sess. 1 (1991) (statement of Senator Metzenbaum). For example, while North Carolina has 4200 farmers, more than a fifth of those in the South who raise chickens and turkeys - accounting for 99% of all broilers and 90% of all turkeys—are grown by farmers under contracts with large poultry corporations. The companies provide the chickens, feed, and medication and the farmers provide the house and equipment. Id. at 320 (Barry Yeoman, Don't Count Your Chickens, SOUTHERN EXPOSURE, Summer 1989 at 21-24).

1978 to 61.5% in 1998. This monopolization of the chicken industry is certainly not unique. According to 1997 figures published by the National Turkey Federation, the top 6 turkey processors controlled more than 50% of the turkey market. *Drover's Journal* (July, 1997) reports that in beef slaughter, the top four beef packers control 81% of the market, and only 5% of the feedlots control over 90% of U.S.-fed cattle. In 1989, five egg production companies controlled almost 20% of the market.[161] These industry giants are extraordinarily powerful and efficient lobbyists, and they are primarily concerned with profit. This profit motive is often the cause of inadequate conditions for animals raised for food or food production.[162]

Consequently, at the heart of this subject lies a simple conflict—the humane treatment of animals versus profit. Farming is a multi-billion dollar industry, fiercely protected by those who gain from it. Many farming practices that increase profits result in increased suffering of animals and it is wrongly perceived that any attempt to treat animals better will result in reduced profits. As Mr. Justice Bell recently noted in the McLibel case in England, "of course, the commercial urge to rear and slaughter as many animals as economically and quickly as possible may lead to cruel practices or a significant number of instances of cruelty in methods designed to be humane, which could be avoided if less attention was paid to profit and more to animals."[163]

Economic pressures are particularly relevant because there is little personal incentive to treat animals raised for food or food production humanely. Unlike many activities, there is minimal cultural pressure to limit cruel practices in intensive farming since very few people are aware of what occurs. This is true even though the products of intensive farming pervade our grocery stores and menus and are uncritically embraced:

> In general we are ignorant of the abuse of living creatures that lies behind the food we eat. Our

---

[161] William D. Hefferman & Douglas H. Constance, Concentration of Agricultural Markets, Department of Rural Sociology, University of Missouri. (May, 1991) (unpublished manuscript, on file with author)

[162] Agribusiness also effectively markets its products. For example, the National Livestock and Meat Board promotes the "Doctor Opinion Correction Campaign," with the aim of "improv[ing] physicians'" attitudes about pork. The campaign is aimed at "primary care physicians in group practices in metropolitan areas and with patient profiles that call for dietary counseling." Over one million dollars is projected to be spent on this campaign. A similar campaign is "Youth Initiative," designed "to provide accurate information about meat to America's young people." NAT'L HOG FARMER, Mar. 15, 1993, at 6.

[163] McLIBEL, Sect. 8.

> purchase is the culmination of a long process, of which all but the end product is delicately screened from our eyes...There is no reason to associate [a neat plastic] package with a living, breathing, walking, suffering animal.[164]

As *The Economist* recently remarked:

> It is all very well to say that individuals must wrestle with their consciences (but only if their consciences are awake and informed). Industrial society, alas, hides animals' suffering. Few people would themselves keep a hen in a shoebox for her egg-laying life; but practically everyone will eat smartly packaged, "farm fresh" eggs from battery hens.[165]

Consumers do not wish to be reminded of the origins of their meat. Thus, meat always "com[es] cooked and reshaped, in a sesame bun or an exotically flavored sauce, as a turkey roll, or as a chicken nugget, in a crumb coating or a vacuum packet, with not a hint of blood in sight."[166] We disguise the source of the meat by describing pig as pork, bacon, or sausage; cow as beef or hamburger; sheep as mutton; calves as veal; and deer as venison. Although meat exists as a living animal prior to the animal's death, our society clearly avoids this fact. In food production, animals are turned into "food-producing units," "protein harvesters," "converting machines," "crops," "grain-consuming animal units" (as defined by the USDA) and "biomachines."[167]

As a result of such economic and societal pressures, the majority of states in the United States have enacted laws mandating that prosecutors, humane enforcement agencies and the judiciary cannot examine farming practices for cruelty or animal abuse once the particular practice is demonstrated to be a customary practice of the United States farming community.

Cruel farming practices should not be excluded from criminal sanction simply because they are accepted by the farming community. Lawmakers should not assume that customary farming practices are not

---

[164]PETER SINGER, ANIMAL LIBERATION: A NEW ETHICS FOR OUR TREATMENT OF ANIMALS 92-93 (1990).
[165]What Humans Owe To Animals, supra note 73, at 12.
[166]Nick Fiddes, Meat: A Natural Symbol, UTNE READER, Mar. - Apr. 1992.
[167]CAROL J. ADAMS, THE SEXUAL POLITICS OF MEAT 68 (1990).

cruel and do not cause unnecessary suffering. Neither should they allow such suffering simply because the practice is one accepted by the farming community. The delegation of power to the farming industry is breathtaking. It is difficult to imagine another non-governmental group possessing such influence over a criminal legal definition; for example, chemical corporations determining that they did not pollute (and, consequently, violate criminal law) so long as they released pollutants in amounts "accepted", or viewed as "customary", by the chemical industry. In effect, state legislators have granted agribusiness a "license" to treat farm animals as they wish.

In the much publicized McLibel case, Mr. Justice Bell concluded that certain common farming practices were cruel. During the trial, McDonalds argued that if a farming practice was the norm it was "thereby acceptable and not to be criticized as cruel." Mr. Justice Bell, however, clearly rejected this argument ("I cannot accept this approach") on the basis that "to do so would be to hand the decision as to what is cruel to the food industry completely, moved as it must be by economic as well as animal welfare considerations."[168] This determination is an unequivocal rejection of the statutory reality in the United States.

The laws governing cruelty to animals either fail to cover the vast majority of domestic animals in this country, are ineffectively enforced, or are painfully inadequate, such as the laws concerning transport time limits and space requirements. If effectively enforced and applied to the practices described in the article, state anticruelty laws that do not exempt customary farming practices could be somewhat effective in preventing institutionalized cruelty to animals raised for food or food production. Unfortunately, the cruel practices described in this article exist in many states where customary farming practices are not exempt from laws protecting animals.[169]

The essential point of this booklet is to demonstrate the absence of a presumed presence of law. We must recognize that our elected representatives are creating a legally protected sphere whereby any act, if it is viewed as customary by the United States farming community, is not found cruel. Thus, even if certain current farming practices are not interpreted as cruel, the enactment of such exemptions creates an arena whereby any future farming practice is possible, no matter how horrific, as long as the practice is a customary one.

The legislatures of the nineteenth century based their anticruelty

---

[168]McLIBEL, Sect. 8.
[169]See Wise, supra note 26, at 207.

laws, in part, on the belief that cruelty to animals hardens the heart of mankind:

> But legislatures err in believing that when they narrow the positive law that codifies this moral precept, as when they exclude farm animals or animals raised under factory-farming conditions from its reach, they modify the moral precept [itself].[170]

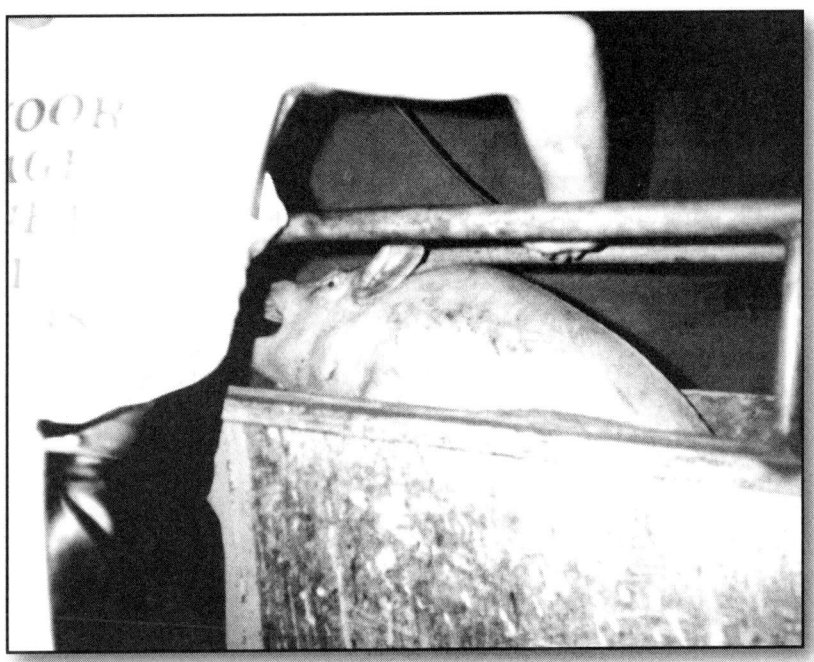

B. The Future?

Since the early nineteenth century, the United States has possessed laws that, on their face, prohibited cruel farming practices. These laws were enacted, in part, because individuals had previously treated animals as they

---

[170]Id. at 209.

Above: A pig being stunned prior to slaughter.

wished. As a result, animals were subject to horrible abuse. Many states recognized that, if left to its own devices, society would exploit animals without regard to moral or ethical considerations. Consequently, states enacted these early statutes to prohibit cruelty to animals such as sheep, pigs, cows, and horses. As time passed, other animals, such as cats and dogs, were brought within the protected sphere of the anticruelty statutes.

Initially, such laws were applied by some prosecutors to certain customary farming practices.[171] However, as time passed, courts and prosecutors became loathe to apply the statutes to common farming techniques. If a court had to determine whether a particular customary farming practice was cruel, it might simply "pass" on the question. As one court stated:

> It must have come to the attention of many that the treatment of "animals" to be used for food while in transit to a stockyard or to a market is sometimes not short of cruel and, in some instances, torturable. Hogs have the nose perforated and a ring placed in it; ears of calves are similarly treated; chickens are crowded into freight cars; codfish is taken out of the waters and thrown into barrels of ice and sold on the market as "live cod"; eels have been known to squirm in the frying pan; and snails, lobsters and crabs are thrown into boiling water...still no one has raised a voice in protest. These practices have been tolerated on the theory, I assume, that, in the cases where these living dull and cold-blooded organisms are for food consumption, the pain, if any, would be classed as "justifiable" and necessary.[172]

In the last 30 years, the process of rearing farm animals has drastically altered; animals are now reared in an industrial and mechanized fashion in order to produce the maximum amount of food,

---

[171] For example, Henry Bergh, the man credited with the creation of the ASPCA and humane enforcement in New York, obtained a successful prosecution under the 1867 New York anticruelty statute for the method by which sheep and calves were transported to the "shambles" (slaughter houses). Similarly, the ASPCA's first series of cruelty enforcement cases dealt with concern about adulterated food for horses and cattle, as well as the transportation of cattle by the railroad. Favre & Tsang, supra note 19, at 14-20.

[172] People ex. rel. Freel v. Downs, 136 N.Y.S. 440, 445 (N.Y. Magis. Ct. 1911) (emphasis added).

i.e., confinement and increased stocking densities.[173] As a result, the consequences of not applying the anticruelty statutes to such farming practices have magnified. Approximately eight billion farm animals are killed every year. The family farm has become a massive machine, abusing animals on a scale not previously imagined. Indeed, most of the abusive farming practices described in the preceding sections are modern developments. Concurrently, certain segments of the public have expressed an increased concern for animal welfare and "animal rights."

In recent years, many state legislatures were no longer content to rely on the courts and prosecutors to interpret anticruelty statutes to exempt cruel common farming practices. This attitude was probably spurred by the notion that "animal rights" philosophies could gain sympathy among prosecutors and judges. It became conceivable that anticruelty statutes might be applied to modern intensive farming techniques, and thus impact the profit-making ability of farms. In response, 18 state legislatures in the last ten years have amended their anticruelty statutes to exempt customary farming practices. In essence, they have handed the farming community the power to decide for itself what constitutes cruelty to animals.

The bizarre result of this trend is that farm animals have been placed in a legal time machine and transported to a time prior to the enactment of anticruelty statutes. Individuals can once again treat animals as they wish. By contrast, European legislatures have responded to modern farming practices by enacting new statutes which specifically prohibit certain cruel farming practices and mandate, at least in part, the proper treatment of farm animals. The contrast between the legal development of Europe and the United States is stark; Europe's statutory development is consistent with the history and the evolution of its anticruelty laws, whereas the United States legal development is one of regression.

If we are to honestly act as a society that condemns the unnecessary suffering of animals, our system of laws must be changed to resemble laws in Western Europe, where certain farming practices have been recognized for what they are: cruel. The current trend in the United States legal system implies we value profit and appetite over any pain felt by an animal. Consequently, animals raised for food or food production do not receive the legal protection from cruelty that other animals receive.

The United States should legislate proper farming practices. In doing

---

[173] ROBERT GARNER, POLITICAL ANIMALS, ANIMAL PROTECTION POLITICS IN BRITAIN AND THE UNITED STATES 44 (1998).

so, the legislature and judiciary can reclaim, from the farming community, the power to define what is cruelty to animals. Given the intense economic pressures from agribusiness in certain states, the legislation should be federal to prevent amendments to state anticruelty statutes that effectively negate the law's effect and result in uneven and inconsistent laws.

For example, the veal crate and the anemic diet for calves, as well as

the confinement of sows in cramped stalls, should be specifically prohibited by statute, as it is in England and other European countries. Similarly, battery cages, force-feeding, beak-trimming, and the killing of chicks by suffocation should be abolished, as should all of the customary practices determined to be cruel by Mr. Justice Bell (listed on page 26). Switzerland, Germany and Sweden have already banned some of these practices. Moreover, no drugs or hormones should be used on farm animals, except to treat disease that is not induced by stressful conditions. The transport of day-old calves should be banned, as in England.

A governmental organization should be created to determine which customary farming practices should be prohibited. This organization should be comprised of farming representatives as well as experts whose

Above: Poultry, unprotected from the elements, is transported in crowded cages stacked high on a truck.

primary concern is the health and welfare of the farm animal. This new body would differ from federal organizations that currently deal with farming issues—the USDA or the Food Safety and Inspection Service (FSIS)—whose concern is for the humans who eat the animals. This organization would also examine all new farming practices to ensure that such practices were humane.[174]

Specific animal welfare measures should also be legislated requiring the provision of adequate fresh water, nutrition for full health and vigor, veterinary care, grazing, shelter, exercise, and housing in compatible social groups under as natural conditions as possible. The housing should not impair the animal's ability to rise, lie down, turn around, groom, and fully spread limbs or wings in any direction. Additionally, the law should mandate anesthetic use in all potentially painful procedures, and it should ban hot-iron branding as well.[175] Once again, Sweden's recent legislation provides a good statutory model.

As stated above, such legislation must be accompanied by effective enforcement and high fines, as is the case with other important criminal laws. A regulatory scheme should exist whereby farms would be periodically open to inspection by an enforcement agency so as to prevent abuse.[176] Likewise, the present transportation and humane slaughter laws should be

---

[174]The Food Safety and Inspection Service (FSIS) presently conducts a similar procedure in relation to slaughter. Section 1904(a) of the Humane Methods of Slaughter Act of 1978 authorizes and directs the Secretary of Agriculture to "conduct, assist, and foster research, investigation, and experimentation to develop and determine methods of slaughter and the handling of livestock in connection with slaughter which are practicable with reference to the speed and scope of slaughtering operations and humane with reference to other existing methods and then current scientific knowledge." 7 U.S.C. § 1904(a) (1994) (emphasis added). Consequently, the FSIS must approve new technologies and procedures in federally inspected plants and determine whether such activities are humane. It is, however, unclear to what extent the FSIS actually incorporates humane considerations in its determination. See, e.g., FOOD SAFETY AND INSPECTION SERVICE, GUIDELINES FOR PREPARING AND SUBMITTING EXPERIMENTAL PROTOCOLS FOR IN-PLANT TRIALS OF NEW TECHNOLOGIES AND PROCEDURES, Directive 10,700.1 (April 11, 1995) which focuses on four areas of concern—product safety, worker safety, environmental safety and inspection procedures—but contains no guidelines expressly relating to humane treatment of animals.

[175]See also, Position Statement on "Factory Farming" (Association of Veterinarians for Animal Rights) 1988.

[176]At present, it seems that there is no active state or federal involvement in the inspection of farms for cruelty to animals raised for food or food production, although certain industries have Industry Quality Assurance Programs whereby the industry regulates itself to assure the quality of the product, i.e., to avoid damaged meat.

properly enforced, and research should be conducted to determine a humane time limit for the transportation of animals raised for food or food production similar to the recent European Commission legislation.

Such a system is necessary given the failure of today's laws to control the cruel realities of intensive farming. The alternative is the present situation with all of its legal and societal inconsistencies. Today, the overwhelming majority of domestic animals in the United States have no real legal protection, yet most people believe that they do. Some animals are protected from cruelty and abuse in this country; others are not. Whether we choose to move toward a more humane and legally consistent future will provide an invaluable postscript to the following observation made by Judge Arnold in 1888:

> [L]aws, and the enforcement or observance of laws for the protection of dumb brutes from cruelty are, in my judgment, among the best evidences of the justice and benevolence of men.[177]

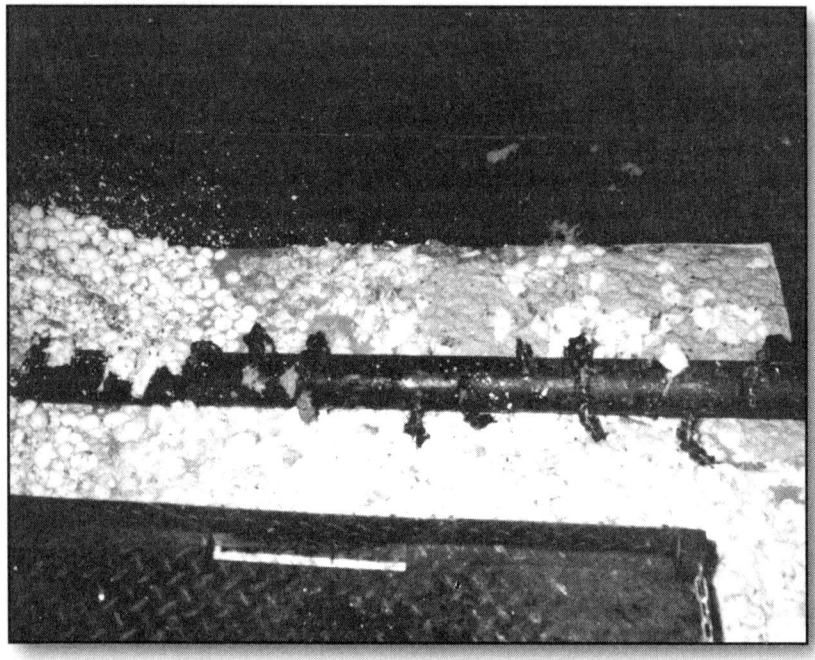

[177] Stephens v. State, 3 So. 458 (Miss. 1888).

Above: Live and dead chicks are dumped in a manure spreader to be used for fertilizer behind a hatchery.

APPENDIX

STATE ANTICRUELTY STATUTES EXEMPTING CUSTOMARY FARMING PRACTICES

ARIZ. REV. STAT. ANN. § 13-2910.03 (1997). The anticruelty statute shall not restrict activities regulated by the Arizona Department of Agriculture.

COLO. REV. STAT. § 18-9-201.5 (1997). The anticruelty statute shall not affect accepted animal husbandry practices. [Amended in 1990]

CONN. GEN. STAT. ANN. § 53-247(b) (1998). Exempts any person who follows "generally accepted agricultural practices" from violating the statute prohibiting the malicious and intentional maiming, mutilating, torturing, wounding or killing of an animal. [Amended in 1995]

IDAHO CODE § 25-3514(5)(9) (1997). The anticruelty statute shall not be construed as interfering with normal or accepted practices of animal identification and animal husbandry and any other activities, practices or procedures normally or commonly considered acceptable. Enforced by the Department of Agriculture. [Amended in 1994]

ILL. ANN. STAT. 510 ILCS 70/13 (1998). The Humane Care for Animals statute shall not affect normal good husbandry practices.

IND. CODE ANN. § 35-46-3-5 (1998). Exempts acceptable farm management practices. [Amended in 1987]

IOWA CODE § 717.2 (1997). Livestock excluded from general anticruelty statute. "Livestock Neglect" and "Livestock Abuse" applies if a person "(a) fails to provide livestock with care consistent with customary animal husbandry practices; (b) deprives livestock of necessary sustenance; (c) injures or destroys livestock by any means which causes pain or suffering in a manner inconsistent with customary animal husbandry practices." In § 717.1(6), "sustenance" is defined as "food, water, or a nutritional formulation customarily used in the production of livestock." [Amended in 1995]

KAN. STAT. ANN. § 21-4310(b)(6) (1997). The anticruelty statute shall not apply to normal or accepted practices of animal husbandry.

LA. REV. STAT. ANN. § 14:102.1(C)(D) (1998). The Louisiana anticruelty statute states that it shall not apply to the "herding of domestic animals", and states for "the purposes of this Section, fowl shall not be defined as animal."

MD. CODE ANN. art. 27, § 59(c) (1997). Exempts customary and normal agricultural husbandry practices including dehorning, castration, docking tails and limit feeding. Also exempts normal human practices to which infliction of pain to an animal is purely incidental and unavoidable.

MICH. STAT. ANN. § 28.245(8)(f) (1997). The anticruelty statute does not prohibit the lawful killing of livestock or generally accepted animal husbandry or farming practices involving livestock. [Amended in 1994]

MO. ANN. STAT. § 578.007(8) (1997). The anticruelty statute shall not apply to normal or accepted practices of animal husbandry.

MONT. CODE ANN. § 45-8-211(4)(b) (1997). Exempts commonly accepted agricultural farming and livestock practices. [Amended in 1991]

NEB. REV. STAT. § 28-1013(7) (1997). Exempts "commonly accepted practices of animal husbandry with respect to farm animals including their transport from one location to another." [Amended in 1990]

NEV. REV. STAT. ANN. § 574.200.(b) (1997). Exempts established methods of animal husbandry, including raising, handling, feeding, housing and transporting. [Amended in 1989]

N.J. STAT. ANN § 4:22-16.1 (1998). Creates a legal presumption that any person who follows the standards for the humane raising, keeping, care, treatment, marketing and sale of domestic animals (to be created by the State Board of Agriculture and the Department of Agriculture, in consultation with the New Jersey Agriculture Experiment Station) shall not violate the anticruelty statute. Such standards were to be enacted by 1996 but as of 1998 have not been promulgated. [Amended in 1995]

N.C. GEN. STAT. § 14-360(c)(2) (1998). The Cruelty to Animals statute shall not apply to "lawful activities conducted for...purposes of production of livestock or poultry." [Amended in 1998]

OHIO REV. CODE ANN. § 959.13 (A)(4) (Anderson 1998). It is unlawful to "keep animals other than cattle, poultry or fowl, swine, sheep or goats in an enclosure without wholesome exercise and a change of air." Subsection (A)(2) states that prior to slaughter, farm animals are also exempted from the requirement for shelter.

OR. REV. STAT. §§ 167.320 (2), 167.315(2) (1997). Exempts good animal husbandry practices from the crime of animal abuse. In § 167.310(2), good animal husbandry is defined according to accepted practices of animal husbandry. Commercially grown poultry and animals subject to good animal husbandry practices are also exempt from (i) the crime of aggravated animal abuse and (ii) requirements of food, shelter, cleanliness, temperature, exercise and space provided for in the animal neglect statute, unless gross negligence can be shown. § 167.335. [Amended in 1985]

Above: Fully conscious chickens in line to have their throats cut.

PA. STAT. ANN. tit. 18, § 5511(c) (1997). Exempts normal agricultural operations. Normal agricultural practices are defined as normal activities, practices and procedures that farmers adopt, use or engage in year after year in the production and preparation for market of poultry and livestock. [Amended in 1984]

S.C. CODE ANN. § 47-1-40(c) (1997). The anticruelty statute does not apply to accepted animal husbandry practices and also specifically excludes fowl. [Amended in 1988 in relation to accepted animal husbandry practices]

S.D. CODIFIED LAWS ANN. § 40-1-33, 40-1-2.4 (1998). Exempts standard and accepted agricultural pursuits and procedures. [Amended in 1991]

TENN. CODE ANN. § 39-14-202(e)(1) (1997). The anticruelty statute does not prohibit "usual and customary practices which are accepted by colleges of agriculture or veterinary medicine." In addition, pursuant to a 1997 amendment, whereas investigations relating to cruelty to non-livestock animals can be conducted by agents of any society for the prevention of cruelty to animals, no entry onto a farm or arrest or interference related to usual and customary agricultural practices can be taken in response to an allegation unless, prior to or at the same time as such entry, arrest or interference, the livestock in question is examined by the "county agricultural extension agent of such county, a graduate of an accredited college of veterinary medicine specializing in livestock practice or a graduate from an accredited college of agriculture with a specialty in livestock." If such an individual does not have probable cause to believe a violation of the anticruelty statute (which exempts usual and customary farming practices) has occurred no action can be taken. TENN. CODE. ANN. § 39-14-211 (1997). [Amended in 1989 and in 1997 in relation to enforcement]

UTAH CODE ANN. § 76-9-301 (11)(b)(ii) (1998). Specifically states that animals within the cruelty statute do not include "animals kept or owned for agricultural purposes in accordance with accepted husbandry practices." [Amended in 1991]

VT. STAT. ANN. tit. 20, § 3901(4) (1998). The Animal Welfare Act does not include "horse, cattle, sheep, goats, swine, and domestic fowl." VT. STAT. ANN. tit. 13, § 352(a)(3) (1998). The Animal Cruelty Law states that livestock and poultry husbandry practices are exempted from the section that deems it illegal to "tie, tether or restrain" an animal in a manner that is inhumane or detrimental to the animal's welfare. [Amended in 1989]

VA. CODE ANN. § 3.1-796.122(C) (1997). Nothing in the anticruelty statute "shall be construed to prohibit the dehorning of cattle." In addition, the time period allowed for a prosecution of cruelty to agricultural animals is one year. By contrast, the time period to prosecute cruelty to other animals is five years.

WASH. REV. CODE ANN. § 16.52.185 (1997). Excludes accepted husbandry practices used in the commercial raising of livestock and poultry.

W. VA. CODE § 61-8-19(e) (1998). Excludes usual and accepted standards of livestock, poultry, and gaming fowl farm production and management. [Amended in 1991]

WIS. STAT. ANN. § 951.14 (1997). Requires the provision of proper shelter for animals but states that nothing in the statute imposes shelter requirements or standards more stringent than normally accepted husbandry practices in the particular *county* where the animal or shelter is located. (Emphasis added).

WYO. STAT. § 6-3-203(f)(v) (1997). The anticruelty statute does not prohibit the use of commonly accepted agricultural and livestock practices on livestock. [Amended in 1994]

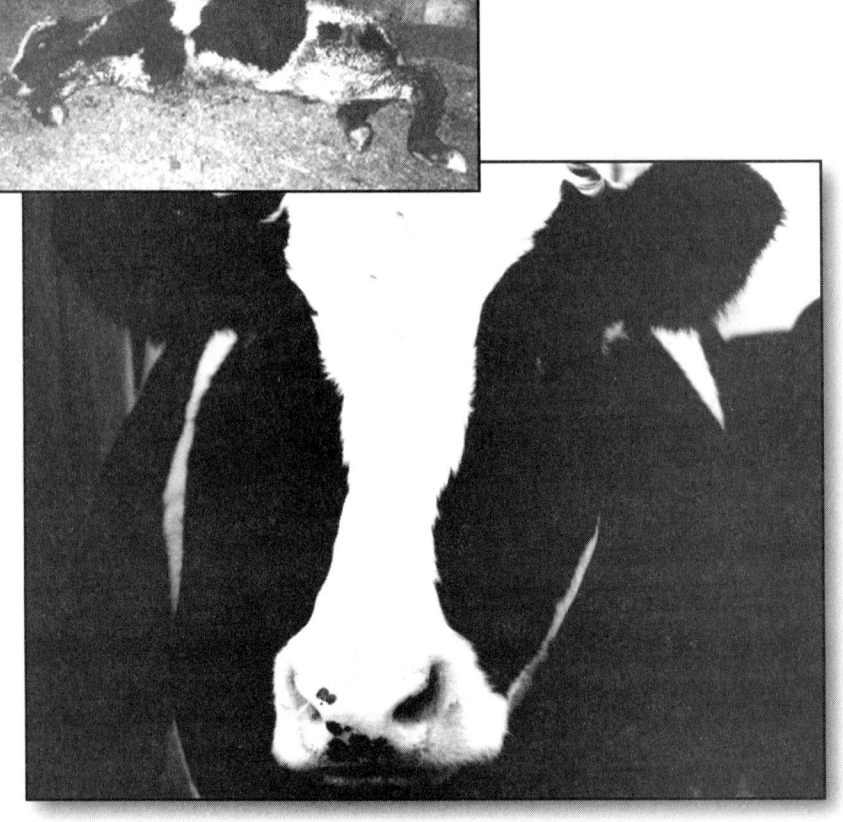

ABOUT FARM SANCTUARY

Farm Sanctuary is a national, nonprofit organization dedicated to rescuing and protecting farm animals. Since Farm Sanctuary began in 1986, we have devoted our resources and time to exposing, and stopping, the cruel practices of the "food animal" industry. And thanks to Farm Sanctuary members, we are changing the way society views and treats farm animals....

• Our unique coast-to-coast shelters provide a "happy

Inset: An incapacitated calf is left for dead at a stockyard. Above: After his rescue, he recovers and is allowed to live out his life in peace at Farm Sanctuary.

ending" for rescued "veal" calves, "stockyard" pigs, and other suffering farm animals. Each year, we rescue, rehabilitate, and provide lifelong care for hundreds of animals rescued from the nightmare of "food animal" production.

- Our research team investigates and exposes abusive practices at slaughterhouses, livestock markets, and factory farms across the country. These efforts are attempting to redefine the way "food animals" are treated under state anti-cruelty laws and setting precedents for humane enforcement efforts for farm animals.

- We actively lobby for passage of federal and state legislation which will ban cruel animal agricultural practices. Our groundbreaking legislative initiatives seek to obtain needed legal protection for animals used for food production.

- We educate millions of people through extensive national news exposes. Farm Sanctuary campaigns and programs have been featured in *The New York Times, The Los Angeles Times, CBS This Morning, National Public Radio, The Wall Street Journal, CNN Larry King Live, USA Today, PrimeTime Live*, and hundreds of other national and regional news stories.

Farm Sanctuary is leading the fight for farm animal rights—thanks to people who care enough to be a part of these efforts. For further information on what YOU can do to help, please contact:

Farm Sanctuary

East	*West*
P.O. Box 150	P.O. Box 1065
Watkins Glen, NY 14891	Orland, CA 95963
Phone: 607-583-2225	Phone: 530-865-4617
Fax: 607-583-2041	Fax: 530-865-4622

E-MAIL: office@farmsanctuary.org
WEBSITE: www.farmsanctuary.org